Practical TPM
Successful Equipment Management at Agilent Technologies

Practical TPM
Successful Equipment Management at Agilent Technologies

James A. Leflar

PRODUCTIVITY
PORTLAND, OREGON

Productivity Press
P.O. Box 13390
Portland, OR   97213-0390
Telephone: (503) 235-0600
Telefax: (503) 235-0909
E-mail: info@productivityinc.com

Managing editor Michael Ryder
Cover design by Stephen Scates
Page design and composition by William H. Brunson Typography Services
Printed and bound by Malloy
Printed in the United States of America

Library of Congress Cataloging-in-Publication Data

Leflar, James A.
    Practical TPM : the method for success at Agilent Technologies / James A. Leflar.
      p. cm.
    Includes index.
    ISBN 0-56327-242-3
    1. Total productive maintenance.   I. Title.
  TS192.L44   2001
  658.2'02—dc21   00-068425

06 05 04 03 02 01   6 5 4 3 2 1

Contents

Preface   xi

Acknowledgements   xiii

PART 1: BASIC TPM PRINCIPLES & CONCEPTS   1

Chapter 1 TPM Fundamentals   3
Six Basic Principles of TPM   5
Continuous Productivity Improvement   7
The TPM Culture Change   9
Principles of Learning   11

Chapter 2 Improving Machine Performance   13
TPM Improves Machine Productivity   15
The Concept of Constraints   16
The Equipment Aging Paradigm   17
Goals of Equipment Maintenance   18
Machine Deterioration and Loss   22
How TPM Attacks Machine Loss   23
The TPM Pyramid of Chronic Conditions   24
TPM Focuses on Machine Component Maintenance   28

Chapter 3 The Team Approach   31
Action Teams   33
The TPM Pilot Team   34
Implementing TPM Action Teams   36
Keys to TPM Success   39

PART 2: STEP-BY-STEP TPM IMPLEMENTATION 41

Chapter 4 Step 1: Restore Equipment to "New" Condition 45
Step 1 Goals 47
Minor Machine Defects 48
    Types of Machine Defects 48
    Component Standards: How Components "Ought to Be" 48
    Cleaning to Detect Machine Defects 51
    Minor Defects Cause Machine Failures 53
    Examples of Minor Equipment Defects 55
    Using Appropriate Team Members to Set Machine Standards 61
Step 1 Toolkit 63
    TPM Program Safety 63
        Safety Management Systems 64
        A Safe Working Environment 64
        Safe Employee Behavior 67
        The "Three Lists" 68
    Cleaning and Inspection Materials 74
    M-Tags 76
    1-Point Lessons 78
    Visual Controls 83
        Visually Controlled Inspection Routes 84
        Visual Route Maps for Frequent PMs 86
        Visually Controlled Production Flow 87
    Team Notebooks and Activity Boards 89
    Cleaning and Inspection Standards 89
Getting TPM Teams Started 95
Advancing Step 1 Team Activities 96
    Cleaning Improvements 97
    Maintenance Access Improvements 98
    Daily Activity Scheduling 101
    Ongoing Cleaning and Inspection Audits 105
Step 1 Master Checklist 105
Step 1 Infrastructure Support 108
Step 1 Deliverables 109

Chapter 5 Step 2: Identify Complete Maintenance Plans 111

Step 2 Goals 113
Identifying a Complete Machine Preventive Maintenance Plan 113
 PM Checklists 114
 PM Schedules 116
  Designing Maintenance Schedules 120
   Graceful Deterioration 120
   Nongraceful Deterioration 121
   Machine Redesigns 123
 Inspection Specifications 125
 Replacement Parts 125
 PM Procedures 125
 Part Logs 126
 Quality Checks 127
Step 2 Master Checklist 129
Step 2 Infrastructure Support 130
Step 2 Deliverables 130

Chapter 6 Step 3: Implement Maintenance Plans with Precision 131

Step 3 Goals 133
On Time and Complete 134
Precision Execution 136
 Basic Technical Skills 136
  Preparing Technical Minicourses 139
  Training Guidelines 143
 Specific Machine Skills 143
 Failure Prevention Skills 144
 Precision Maintenance Support Tools 144
  Precision Documentation 144
  1-Point Lessons 152
  Time to Learn 153
  A "Partner and Practice" System 154
  An Organized Workplace 155
   Creating a Visual Workplace 167
  Spare Part and Tool Management 160
   Open Part Stock 160
   Controlled Part Stock 164
   Maintenance Tools 166

Elevating Knowledge, Skills, and Maintenance Roles 170
    Elevating the Operator's Role 170
    Elevating the Engineer's Role 171
Step 3 Master Checklist 171
Step 3 Infrastructure Support 173
Step 3 Deliverables 173

Chapter 7 Step 4: Prevent Recurring Machine Failures 175
Step 4 Goals 177
PM Evaluations 178
    Rebuild and Swap Technique 178
Failure Analysis 179
    Failure-Prevention Trouble Lists 180
        Start-up Level: Machine Failures with Known Repairs 183
        Practiced Level: Recurring Machine Failures
        with No Known Repair 183
        Advanced Level: All Machine Failures 183
    Failure-Analysis Tools 184
        5-Why Analysis 184
        Guessing Wastes Resources 188
        Maintenance Analysis 189
        P-M Analysis 191
        P-M Lite 192
        Physical Analysis 196
        PDCA 201
        Countermeasure Plans 205
Step 4 Master Checklist 206
Step 4 Infrastructure Support 207
Step 4 Deliverables 207

Chapter 8 Step 5: Improve Machine Productivity 209
Step 5 Goals 211
Lubrication Analysis 212
Calibration and Adjustment Analysis 216
Quality Maintenance Analysis 218
Machine-Part Analysis 222
Condition-of-Use and Life Analysis 222

Productivity Analysis: The Six Big Productivity Losses   228
    Machine Breakdowns   229
    Machine Setup and Adjustment Time   229
    Product Scrap   231
    Low Product Yields   232
    Minor Machine Stoppages   232
    Reduced Machine Speed   233
Extended Condition Monitoring   234
    Vibration Analysis   235
    Ultrasonic Analysis   236
    Wear Particle Analysis   236
    Infrared Thermography   237
    Video-Imaging Analysis   238
    Water-Quality Analysis   238
    Motor-Condition Analysis   239
    Jigs, Fixtures, and Test Gauges   240
    Continuous Condition Monitoring   241
Maintenance Cost Analysis   244
Step 5 Master Checklist   245
Step 5 Infrastructure Support   246
Step 5 Deliverables   246
Summary of TPM Steps   247

PART 3: LEADING THE CHANGE   249

Chapter 9 Vision, Strategy, and Tactical Plan   251
Agilent's Manufacturing Vision   254
Agilent's Manufacturing Strategy   254
Agilent's Manufacturing Tactical Plan   254

Chapter 10 Focusing Machine Improvement Activities   255
The Pareto Approach   258
The TPM Approach   260
    Focusing TPM Steps   261

Chapter 11 Improvement Metrics   265

Overall Equipment Effectiveness (OEE)   268

Total Production Ratio   270

    TPR Derivation Terms   271

Number of Machine Failures   274

Duration of Machine Failures   274

Machine Failure Histogram   275

Machine Capacity Assurance   276

Linear Product Flow   281

Factory Output and Product Cost   285

Chapter 12 Managing New Behaviors   287

The ABCs of Behavior   289

    Behavior Antecedents   290

    Behavior Consequences   290

      SSIP Rules   291

Celebrations   293

Shaping Behavior   293

Performance Improvement Process   293

Chapter 13 Team Activity Boards   295

Chapter 14 Sustaining TPM Changes   299

TPM Gap Analysis   302

Chapter 15 Final Thoughts   303

Appendix A Example of Maintenance Training Material   309

Appendix B Answers to Questions about Fasteners   333

Glossary of Agilent Manufacturing Terms   341

Index   343

Preface

In late 1999, Hewlett-Packard began the process of splitting itself into two companies. One company now comprises the computer product lines; the other comprises the test and measurement product lines. The computer business retained the name Hewlett-Packard because of its retail recognition. The test and measurement division—Hewlett-Packard's original business—became Agilent Technologies. The Hewlett-Packard plant, where all of the Total Productive Maintenance (TPM) activities described in this book were created, is now part of Agilent Technologies.

Agilent Technologies operates an integrated circuit fabrication plant (IC fab) in Fort Collins, Colorado. Despite splitting from the computer division, our people, our business, and the TPM activities begun in this plant remain the same. Only the name on the building has changed. Throughout the remainder of this book, the IC fab will be referred to as an Agilent IC fab, despite the fact that it was a Hewlett-Packard fab while the TPM program described in this book was developed.

Agilent's organization consists primarily of production operators, maintenance technicians, and engineers. Operators are primarily responsible for moving product through the factory floor. Technicians maintain equipment in proper working order. Engineers resolve many chronic production problems with improved manufacturing processes and equipment designs, and also develop new products and technologies.

Please keep this structure in mind: the TPM methods described in this book were tailored to Agilent's organization as it existed when its TPM program got underway. This is not to say that every company implementing TPM must have this type of organization. TPM is primarily a way of using your own organization to solve your own manufacturing problems, but its implementation does not require you to reorganize your employee structure.

In many factories, such as automobile assembly plants, there is a strong interaction among operators, their machines, and the manufactured product. In these situations, the condition of the machines and the quality of the products are usually apparent to the machine operators. This is not true in an IC fab. Product wafers are most often processed inside of machines—many of them inside vacuum chambers—completely out of view of the operator. Many machines are bulk-headed into equipment chases for contamination control so that no part of the machine is even visible to the operator, except the wafer load and unload stations.

The quality of the IC manufacturing process is equally difficult for operators to assess. In some cases, quality feedback indicating machine performance is not obtainable until weeks after a machine processes the wafers. Because of this distant relationship among operators, their equipment, and the manufactured product, maintaining optimal machine conditions at all times is extremely critical in the IC industry. Essentially, we must blindly control the inputs to each process step—the machine conditions—to be assured of a high-quality finished product. This is one of the unique challenges of being in the integrated circuit business. This is also one of the major strengths of TPM.

The TPM principles presented in Part 1 of this book are universal. Only the implementation details must be tailored to each user. Part 2 provides descriptions of Agilent's five-step TPM program. It includes many examples from Agilent's own experience to illustrate these TPM steps. However, the TPM methods described in this book are not limited to use in the IC industry. Nuts, bolts, and other machine components don't know in which industry they are being used. Once the TPM principles and step-by-step instructions are understood, they can be applied to any machine, organization, or industry.

People are the key to solving productivity problems. Automation and new technology alone will not get a company to world-class levels of factory productivity; people still need the leadership and tools to deliver higher productivity results. TPM methods provide a step-by-step approach to meet these needs.

Agilent Technologies

Innovating the HP Way

HEWLETT®
PACKARD

Acknowledgements

Agilent began investigating Total Productive Maintenance (TPM) methods in 1995 in an attempt to improve business results by improving factory floor productivity. After a thorough investigation and pilot program, TPM implementation began throughout the plant early in 1997.

Our success so far could not have been accomplished without the help of many people. I would like to recognize a few of those who helped develop TPM in our manufacturing operation.

Mr. Masaji Tajiri, an independent consultant on TPM and other manufacturing methods in Japan, was Agilent's principal productivity advisor and visited our site time and time again over the years, guiding our TPM progress one step at a time. He is largely responsible for shaping Agilent's TPM program into what it has become today.

Mr. Al Barahni, TPM Program Manager at Procter & Gamble, first introduced me to TPM concepts. Mr. Atsuhiro Hayakawa, TPM Manager; Ms. Judy Eddy, Assistant Manager of Assembly Production and TPM; and Mr. Pete Caldwell, Maintenance Manager at Michigan Automotive Compressor, Inc., showed me first-hand what TPM methods could accomplish on a complex manufacturing line. Japanese Institute of Plant Maintenance (JIPM) consultants taught me many TPM skills in numerous classes that I attended. I also attended many TPM events sponsored by Productivity, Inc., and always learned something new. Ms. Betty Vanderlin, Senior Operations Manager for Silicon Products at Motorola, offered me many insights into creating a successful TPM implementation. Mr. V. A. Ames, TPM Program Manager at Sematech (an organization of cooperating semiconductor manufacturers) sponsored many meetings and programs in which I participated.

Primarily, though, the TPM steering committee is responsible for the successful deployment of TPM activities throughout our IC fab. Our steering committee consists of Mark Anderson, IC Fab Manager, who took the first initiative to implement TPM; Mary Verhow, Production Manager; John Vaughn, Maintenance Manager; Stan Strathman, Engineering Manager; Tracy Ireland,

Factory Development Manager; Lee Severson, Production TPM Coordinator; and Chad Haight, Training and Documentation Group Manager.

Mr. Kevin Funk, Senior Factory Development Engineer, spent hundreds of hours working to create numerous infrastructures that supported our TPM development. He also helped develop many of the implementation strategies that kept our TPM program moving forward.

Jerry Akers, Tim Hailey, Karen Williams, and John Williams—all programming specialists in Agilent's Information Technology department—wrote and modified all of the software used to integrate TPM activities into our business operating systems.

Ms. Catherine Varra-Nelson, a principal in the TPM office during the early years of our TPM activities, supported the creation of our first TPM teams, which focused on creating clean, defect-free equipment throughout the IC fab. Working with Catherine were many production operators acting as team coordinators—too many to mention by name—who were the real heroes of our early success at reducing equipment failures.

I need to recognize, most of all, the core members of our TPM pilot team—Mike Verville, Ed Hockensmith, and Travis Swearingen—who helped to develop almost all of the TPM activities described in this book. They put in a tremendous effort, including working on their days off, to complete a successful TPM pilot program. They were a creative, open-minded, hardworking, high-performance team. Agilent's TPM program could not have been developed without their contributions.

Agilent's TPM Pilot Team. From left to right are Jim Leflar; Mike Verville and Ed Hockensmith, Equipment Maintenance Technicians; and Travis Swearingen, IC Process Engineer

I would also like to thank Linda Gallagher and Alice Kober of TechCom Plus in Denver, Colorado, for helping me edit and format the manuscript. Thanks also to Jim Anderson and Todd Cito of Wind River Technologies for providing us with machine design services for all of our TPM activities.

I learn more about TPM methods and implementation every day, and I could have continued to develop this book indefinitely. Thus, as it stands, *Practical TPM: Successful Equipment Management at Agilent Technologies* is a document of the process that Agilent has learned from its TPM journey so far; by necessity, it excludes many TPM practices that are still beyond our horizon. Therefore, the book might be described as the ABCs of TPM implementation rather than the A through Zs.

All of the examples, anecdotes, and data used throughout the book are real, taken from my own personal experiences both at Agilent and elsewhere. Some of this material has been fully implemented in Agilent's IC fab. Some has only been piloted and is in the process of being deployed throughout the plant.

The opinions expressed in this book about TPM implementation are entirely my own. Any deficiencies brought to light by the publishing of this book are mine alone and should not reflect poorly on my teachers or co-workers.

James A. Leflar

PART 1

Basic TPM Principles & Concepts

Total Productive Maintenance (TPM) is a proven methodology used to increase machine productivity. To implement TPM effectively, it is important to first understand several underlying principles and concepts.

In Part 1, the following essentials are addressed:

- **TPM Fundamentals**
- **Improving Machine Performance**
- **The Team Approach**

1

TPM Fundamentals

What is TPM?

How is it used to increase productivity?

What types of culture changes are required to
implement TPM successfully?

SIX BASIC PRINCIPLES OF TPM

TPM activities are based on six simple principles that improve equipment productivity. Note that these principles are often the opposite of conventional manufacturing wisdom.

1. Minor defects are the root cause of most equipment failures and must be completely eliminated from all equipment. Equipment with minor defects will always find new ways to fail, and improvement activity will never be able to keep pace with the failure rates of the machine.

   We call this the "Roseannadana[1] Syndrome." As *Saturday Night Live* character Roseanne Roseannadana would say, "It just goes to show you, it's always something. If it's not one thing, it's another." Eliminating minor defects in machines will prevent many machine failures.

   Some minor defects are caused by machine deterioration; others are man-made. Man-made defects are caused by people who do not have the proper knowledge, skill, tools, or parts to perform their assigned machine tasks. One of TPM's major activities is relentlessly exposing, correcting, and preventing minor equipment defects.

2. Properly planned maintenance routines can prevent almost all sporadic equipment failure. In Agilent's manufacturing terms, well-maintained machines make many wafers. Poorly maintained machines make fewer

---

[1]Roseanne Roseannadana was a fictional character portrayed by the late commedienne Gilda Radner.

wafers. Broken-down machines make none. Scheduled maintenance is the foundation for all other TPM activities.

3. Cross-departmental teams can advance equipment performance with much greater ease than efforts made by any single department working alone. This is especially true for chronic failures and quality problems. Departments working independently will not produce world-class results. Nor can the task of improving machine productivity be placed entirely in the hands of the maintenance department.

4. Continuous learning is at the heart of continuous machine improvement. Machines do only what people make them do—right or wrong—and can only perform better if the people taking care of them acquire new knowledge and skills regarding equipment care. TPM activities always have two aims: achieve optimal equipment conditions, and achieve optimal human performance. These are inseparable because human performance creates the equipment conditions.

    TPM elevates human skills and the role people play in equipment maintenance and productivity improvement. This concept is illustrated in Figures 1-1 through 1-3.

5. Machines with effective preventive maintenance programs make more product than machines that are only repaired when they break down. TPM programs and improved maintenance plans are often avoided because people believe that machine PMs reduce the time that the machines can run in production. This TPM principle is hard for people to grasp. It is easier to

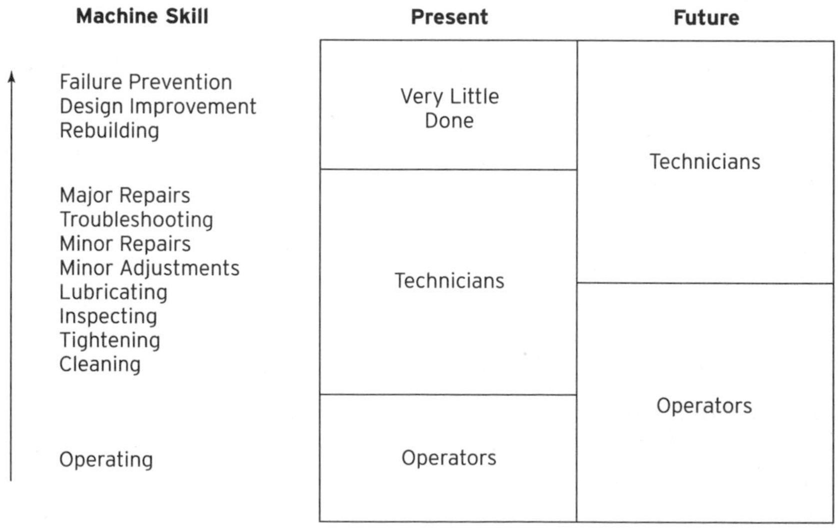

Figure 1-1. As operators and technicians acquire new skills, they can elevate their role in equipment care. Improved equipment care translates into improved equipment performance

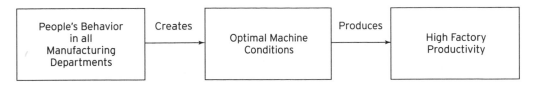

Figure 1-2. Improved behavior on the factory floor translates directly into improving machine performance

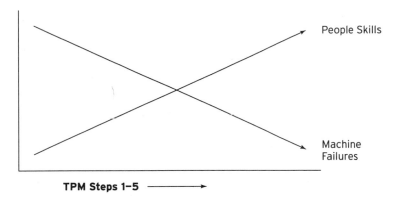

Figure 1-3. The development of people skills directly affects machine performance

believe that more products will be made if machines are kept running for as long as they can and are quickly repaired when they break down.

6. Effective preventive maintenance plans require less technician time than the time required to repair poorly maintained machines. This is contrary to the common belief that the more comprehensive a preventive maintenance plan becomes, the more technicians will be required to maintain equipment.

Unfortunately, many manufacturing managers think about maintenance the way that most homeowners think about a water heater. They have to have it but don't want to spend one penny more on it than is absolutely required. This belief causes many factory managers to limit maintenance plans for equipment to their absolute minimum.

CONTINUOUS PRODUCTIVITY IMPROVEMENT

All factories, no matter what their final product, produce the same output in terms of manufacturing metrics. We usually describe these metrics as deliverables to our customers—TQRDC:

- Technology
- Quality

- Responsiveness
- Delivery
- Cost

If these results are satisfactory to our customers, we can turn them into cash sales and, hopefully, profits—the real product of our manufacturing business. However, in today's rapidly changing business environment, we cannot focus only on profitability. We must continually improve in all five business metrics. Our competitors improve quickly. We must do so as well.

Most manufacturing employees come to work every day to carry out their production or support routines. In Agilent's IC fab, production operators move lots, technicians repair equipment and perform equipment PMs, and engineers support the manufacturing processes carried out on the production floor in many different ways. We generally refer to these routines as "turning the crank."

However, part of any company's daily routine should also include delivering continuous improvements to the company's bottom line; TPM activities are one way to do this. TPM's step-by-step methods get all employees involved in improving their work areas and equipment as part of their daily job routine. Improving factory productivity should not be an extracurricular activity, nor can it be left to just a few individuals.

However, continuous improvement by itself is not sufficient for continued business success. Continuous improvement activities can only produce the desired results if they meet two criteria:

1. We must continually improve our business results *at a competitive pace*. We don't have to have a perfect business, but we do have to keep up with or ahead of our competition. If our performance is not currently competitive, then we will have to improve at an even faster pace.

2. We must focus our improvement activities on the vital few instead of the trivial many. For instance, safety, product shipments, wafer cost, and new business are part of Agilent's vital few business interests today. Expending our limited and valuable resources on improvements that do not contribute to these interests is wasting these resources to some degree. Employees must be made to focus their improvement activities on the most important business results. Without an active attempt to focus all employee improvement activities on system goals, these improvement efforts will largely be aimed at improving only local goals. TPM activities provide a proper improvement focus.

One of the most important metrics for any manufacturer is product cost. One of the easiest ways to reduce manufacturing costs of a product, especially one produced by expensive equipment, is to maximize the productivity of the machines that produce the product.

A company cannot make business gains solely by using cost-cutting measures because it cannot cut costs enough to become a world-class competitor. Instead, it must invest resources in productivity improvement. This generally increases factory throughput and cuts costs at the same time.

THE TPM CULTURE CHANGE

TPM implementation changes the maintenance culture of an organization in many ways. However, this need for a change in people's thinking makes successful TPM implementation difficult. People naturally resist change. They want to think of TPM only as a project tool, instead of as an ongoing process that becomes part of their normal work routine. Table 1-1 indicates some of the culture changes that successful TPM implementation brings about.

Any human skill can always be improved to a higher standard. In fact, skills are best developed in elevating stages. It is not feasible to expect people to achieve a high level of performance when first undertaking a new activity. For instance, when a karate student first learns to perform a front kick, the kick has many flaws.

Table 1-1.

Old Culture	New Culture Created with TPM Activities
Only the top few Pareto problems are resolved, using any means possible to make improvements.	All minor defects in a machine are eliminated. Machine performance is continually improved with the methodical and repeated application of TPM steps.
Improvement methods are implemented by individuals or teams in any way that they see fit.	Improvement methods are rigorously defined and are expected to be implemented precisely.
Improvements in the organization's work methods and processes are localized by each team as they desire.	Improvements in the organization's work methods and processes are coordinated by managers, so the entire organization is learning and benefiting from improved techniques. Even improvement methods themselves are continually being improved.
Machine problems are resolved one at a time, reactively. Ultimately, improvements only occur in systems that have failed.	A reliable and systematic improvement process is applied to a machine to address all productivity losses proactively. Failures are prevented before they occur.
Only results are measured by managers.	Both results and the process used to obtain the results are measured by managers.
Improvement steps are taken as absolute—once completed they are not revisited.	Improvement steps are revisited as people's skills improve, and expectations for their performance are raised.

But once that student achieves a certain level of performance, he or she can advance from a white belt to a yellow belt. However, more practice will be needed for an even higher-level belt. Once the student has sufficiently improved again, the front kick skill will next qualify the student for an orange belt, and so on.

TPM activities are improved in a similar fashion. A team first learning what it means to create a "clean and defect-free machine" will do so at a lower level than a team that has more experience working with TPM activities. But as teams advance their skills, they can be held to a higher standard for "clean and defect-free." This first TPM step can be continually revisited and improved, even as teams have moved beyond this step and are implementing Steps 4 and 5. Improvements in TPM methods and in people's skills never cease.

Eventually, as the way people work on the factory floor changes, the culture change becomes observable. For example, the following changes in the way maintenance technicians spend their time might be observed after TPM activities have progressed.

Maintenance Technician Activity	Before TPM	After TPM Progress
Machine Repairs	70%	10%
Machine PMs	20%	40%
Improvement Activities	10%	50%

People typically voice their resistance to changing the maintenance culture through TPM with comments like these:

- "We don't understand TPM activities. They don't make any sense."
- "We don't have time to implement this 'extra' work when we are so busy repairing broken machines."
- "We don't have enough scheduled machine downtime to carry out TPM activities."

The truth is that investing people's time in TPM activities not only improves machine performance, but actually *reduces* the amount of work people must do to keep their machines running well. TPM activities eliminate a great deal of traditional work and replace it with new kinds of work that actually take less time. This is a win-win situation for both people and equipment. Machines run better with less maintenance time than was required with breakdown maintenance.

But these results are achieved only by people performing at high levels. Elevating people's knowledge and skill is the key to making TPM activities successful and achieving improved factory productivity. This elevation begins with the formation of the first TPM teams and the initiation of Step 1 activities.

TPM activities bring about important changes on the factory floor.

From	To
An overwhelming maintenance workload	A manageable maintenance workload
Many equipment failures	Very few equipment failures
A marginally skilled workforce that carries out its daily routine with little learning	A highly skilled workforce engaged in continuous improvement activity

PRINCIPLES OF LEARNING

Agilent employs a simple model of how people learn that we call LUTI, which stands for Learn, Use, Teach, and Inspect.

- **Learn:** People are first introduced to new knowledge by teaching. The teaching can be conducted by written word, explanation, demonstration, and so on. At this point, a person generally only gains a rudimentary academic knowledge of the subject.
- **Use:** When people practice what they are learning—especially on something of value to them instead of "example" situations—they begin to acquire a true understanding of the subject and begin to develop proficiency. "Use" takes people to a higher level of skill than they acquired in their introductory learning experience.
- **Teach:** When people teach what they have learned to others, they find that they learn more about the subject themselves. It takes a higher level of understanding to teach what they are doing to others than to simply do it themselves.
- **Inspect:** Once people reach a plateau of skill with a subject, they need to inspect their position on the scale of knowledge and see how they can advance to a higher level. Once the higher-level knowledge is identified, they begin by going back to the "learning" step for this new knowledge. Thus, the LUTI cycle is repeated as people's knowledge and skill "spiral" to higher and higher levels.

Another learning concept is that people learn best when subjects are taught to them in the following ways:

- **Break a large subject down into small pieces.** Organize the pieces so that one step builds on the knowledge acquired in the previous step. (Imagine learning mathematics any other way!)
- **Learn by doing.** People must actually apply what they are learning about to truly understand it. People do not gain a high level of understanding of a subject just by reading about it or seeing it demonstrated.
- **Provide guided practice.** As people learn by doing, a knowledgeable instructor should be available to reinforce what they are doing correctly and guide them to improved levels of understanding as they progress.

2

Improving Machine Performance

How does TPM improve machine productivity?

How is it possible to save time and money by improving maintenance technology?

TPM IMPROVES MACHINE PRODUCTIVITY

Like many manufacturers, making everything from computer chips to potato chips, Agilent uses machines to manufacture its products. The fact is machines do virtually 100 percent of the product manufacturing work. The only thing we people do, whether we're operators, technicians, engineers, or managers, is tend to the needs of the machines in one way or another. The better our machines run, the more productive our shop floor, and the more successful our business.

TPM is an equipment-focused improvement effort; we work on creating the ideal equipment state. Any gap between our current state and the ideal state must be closed. These gaps are created in part by: deficiencies in our equipment maintenance plan; people's lack of knowledge as to how to perform their work correctly; and weaknesses in machine, process, and product designs.

In order to change equipment performance, operators, technicians, and engineers must change their own mindsets and work habits. They must learn to tackle improvement issues together as a team rather than separately. They must adopt new mindsets with each TPM step. These changes in people's thinking and behavior improve equipment productivity.

We may never achieve the ideal state for all of these factors on our shop floor, but we are going to move toward them as fast as we can and do so as long as we are in business. Agilent does benchmark other IC fabs in our industry, but our focus is to do more than keep up with our competitors. Our goal is to achieve the ideal factory state—zero losses, or at least world-class levels of machine productivity in our industry.

The first goal of TPM is to prevent equipment failure. Every machine should be kept running as well every day as it has on its very best day. This is primarily achieved by developing a sound maintenance regimen and continually restoring machine deterioration to keep the performance of the machine consistent from one day to the next. Once this is achieved, other productivity losses can be eliminated to make the machine run better than it ever has before.

A TPM team at Agilent recently attacked productivity losses on a very complex cluster tool used to deposit a layer of oxide on IC wafers. The operators, equipment techs, process techs, and engineers on this team used the TPM methods described in this book to improve this machine's productivity. In less than nine months, the machine's capacity was increased by over 50 percent without a single machine design change. The performance improvement of this machine is displayed in Figure 2-1.

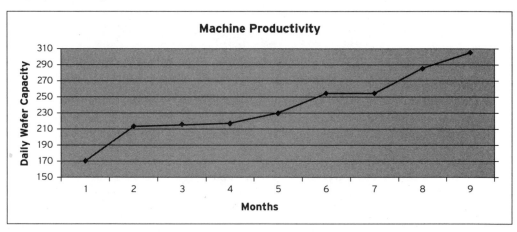

Figure 2-1. Productivity results achieved on a complex machine using TPM improvement methods

THE CONCEPT OF CONSTRAINTS

There are typically two types of machine classifications on any manufacturing floor: bottleneck equipment and non-bottleneck equipment. Bottleneck machines are those machines or groups of machines that have the lowest manufacturing capacity in the factory. Non-bottleneck machines are the remaining machines, which have a higher throughput capacity than the bottleneck machines.

Almost all factories have both types of machines, because it is virtually impossible to balance equipment capacities in any manufacturing operation. There is always a theoretical throughput limiter, even if it is external to the manufacturing floor, such as sales below the factory's capacity.

Machines operate very independently in Agilent's IC fab—they are not tied together in manufacturing lines, as are machines in many other industries. If our operation requires 3½ of a certain machine, we, of course, own 4 of these machines, making them non-bottleneck equipment. For these machines, reliability is more important than capacity.

Improving productivity involves improving the performance of both bottleneck and non-bottleneck machines alike. The reason for bottleneck productivity improvement is obvious. Since these machines have the least capacity of all factory equipment, improving bottleneck productivity improves total factory capacity, in essence increasing the utilization of every piece of equipment in the factory.

Improving non-bottleneck equipment performance is also important because these machines feed the bottleneck. If they stop producing product for too long a time, the bottleneck will starve and also stop production. This production is lost forever, because the bottleneck cannot recover lost production time unless it is run on overtime. Of course, overtime raises manufacturing costs. If machines run 24 hours per day, seven days a week, as they do in Agilent's IC fab, there is no way to recover lost bottleneck production.

To the surprise of many people, most factories are not limited in their production output by their bottleneck machines. Evaluation of data from many factories indicates that most produce less than the capacity of their designated bottleneck machines. Most factory output is actually limited by reliability problems in *all* machines, *not* the bottleneck machine's theoretical capacity. This is why TPM improvement activities must be applied to all equipment, not just the designated factory bottleneck. Improving the reliability of all factory equipment is required to improve overall factory productivity.

 Minimizing manufacturing costs is achieved primarily by improving the productivity of all the machines on the factory floor, both bottleneck and non-bottleneck machines alike. This is a daunting challenge unless the entire organization pitches in and applies proven methods to the task.

THE EQUIPMENT AGING PARADIGM

Most people share a common belief that a new machine is "the best that it will ever be" and that it will continually deteriorate into a worse state as it is used in production. At some point it will become so deteriorated that it will need to be replaced with a new machine.

TPM implementation creates an opposite attitude about equipment aging—that a new machine is "the worst that it will ever be." The more we operate and maintain a piece of equipment, the more we learn about it. We use this knowledge to continually improve our maintenance plan and the productivity of the machine. We would only choose to replace a machine should its technology become obsolete, not because it has deteriorated into a poorly performing machine.

The last day of a machine's use on the factory floor should be its best performing day ever. These attitudes about equipment performance are illustrated in Figures 2-2 and 2-3.

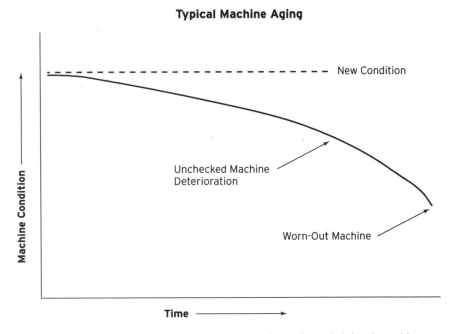

Figure 2-2. Aging machine performance of poorly maintained machines

GOALS OF EQUIPMENT MAINTENANCE

There are two basic goals for equipment maintenance:

1. **Condition maintenance** maintains proper machine conditions so that all components live a natural lifetime.
2. **Replacement maintenance** replaces or services components at the end of their life, but before they fail.

The concept of condition maintenance is illustrated in Figure 2-4.

Components often seem to have unpredictable lifetimes, as shown in failure curve (I), but their life expectancy is "unpredictable" only because they are not being properly maintained. Once they are properly maintained, the average life expectancy of the components will usually increase considerably, and the distribution of the components' life expectancy will be much tighter.

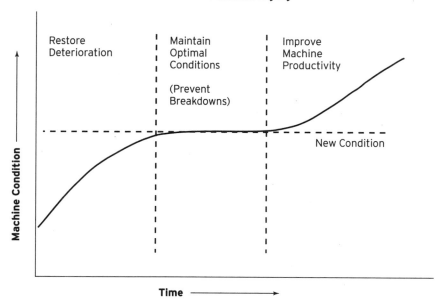

Figure 2-3. Aging machine performance with TPM methods applied

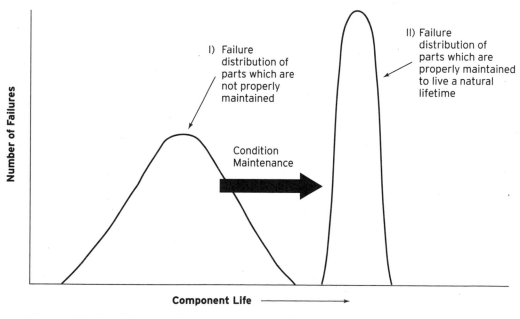

Figure 2-4. Condition maintenance keeps parts in useful service for a natural lifetime

Improper conditions that cause early failures in machine components can originate from numerous causes. True condition maintenance attacks all the following causes of "unpredictable" component lives.

1. Weak machine design by design engineers:
   - Components too undersized to handle the normal stresses applied to them, such as a chain transmitting more power than it was designed to handle.
   - Components used at the extremes of their design range, such as a 1000 sccm mass flow controller being used to control a gas flow of 5 sccm. (These devices work best at 20 to 80 percent of their design rating.)
2. Misapplication of a machine for its intended purpose by application engineers:
   - A 70-ton press is specified for an operation requiring 75 tons of force.
3. Mistreatment of the machine by the operators and technicians who use it and care for it everyday:
   - Slamming doors and similar rough handling of machine components by operators.
   - "Baleing wire" and "chewing gum" repairs made by technicians, vice grips used on hex nuts, and so on.
4. Underdeveloped maintenance plans that do not maintain the required conditions-of-use for machine components:
   - Minor defects neglected in machines often cause accelerated deterioration in nearby components, shortening their lives in unpredictable ways.

If the life of a component is not predictable, specifying a maintenance interval is quite difficult. Choosing service interval (I) as shown in Figure 2-5 would mean replacing most parts long before they were worn, increasing machine downtime for high-frequency maintenance activities and consuming additional technician resources.

Choosing service interval (II)—the average life of the part—would mean half of the parts would fail before they were replaced. As long as the component life is so unpredictable, no scheduled maintenance plan is useful. Even condition-based maintenance plans for these components cause an excess amount of maintenance work and cost. The goal is to improve machine performance with less maintenance work, not more.[1]

Only with proper condition maintenance can many parts achieve a reasonably predictable life expectancy. Once that is achieved, replacement maintenance plans also need to be developed to replace parts as they approach the end of their natural lives. These maintenance plans can be time-based, use-based, or

---

[1] More details about achieving natural component lifetimes can be found in other sections of this book: on page 192 and on page 222.

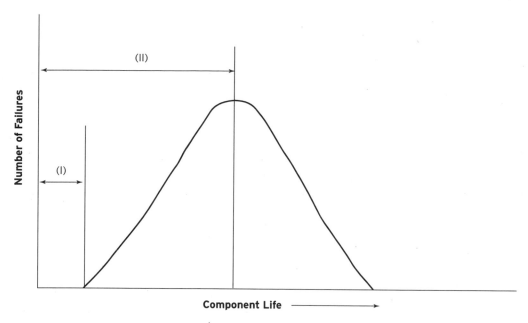

Figure 2-5. Parts with a wide range of life expectancy cannot have a good replacement interval

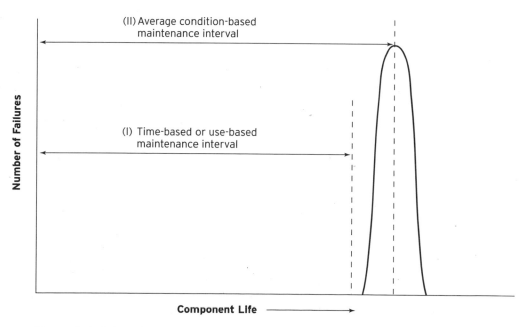

Figure 2-6. Scheduling replacement maintenance for parts with predictable life expectancies

condition-based. Maintenance intervals for these plans are shown as intervals (I) or (II) in Figure 2-6.

MACHINE DETERIORATION AND LOSS

A primary focus of TPM is to evolve a broad range of preventive maintenance plans for equipment. The reason for this is that most of our machines have already demonstrated that on any given day they can do the job they are supposed to do, but they don't always do this job well.

For example, Agilent had a three-year-old wafer-handling robot that ran very well for two years. In fact, there was one three-month period during that time when it ran without a single error. Clearly this wafer handler was designed to do the job we needed done. In its third year, however, the robot ran with numerous errors and stopped the machine from running almost every single day. This kind of repeating failure is known as a *sporadic failure*. Sporadic failures are deviations from the machine's "normal" performance.

What causes such a repeating sporadic failure? In this case, the wafer-handling system deteriorated from years of continuous use. After all, if a machine is running, it is deteriorating in some way. As it deteriorates, its performance varies. So maintenance—the continual restoration of deterioration—makes a machine continue to run every day like it did on its very best day to date. (Assuming, of course, that maintenance procedures are implemented correctly and do not contribute additional minor defects to the machine.)

However, none of our machines has probably seen its best possible day. Our equipment contains productivity losses that are hidden from our view. If we were to discover these hidden productivity losses and eliminate them, the equipment could run better than it ever has before. These continual losses are called *chronic losses* and are most often thought of as the design limit of the machine. However, in most cases, chronic losses have a variety of real causes, which TPM activities can eliminate. The true design limit of the machine probably has a much lower level of loss than most people believe. Even design losses can often be reduced with simple design improvements to the machine. All of these loss concepts are illustrated in Figure 2-7.

Chronic loss is the "normal" operating state of the machine. Chronic losses are generally not repaired, as they are not even considered losses; they're just seen as the way the machines are. Some chronic losses—for example, a certain regular minor stoppage—might simply be reset and the machine operation continued, with nothing ever found wrong with the machine or repaired.

Sporadic loss is a sudden departure of the machine from its "normal" operating state. Equipment that experiences sporadic machine failure is typically returned to production service by troubleshooting and repair work.

Many sporadic losses are caused by one of the two types of machine deterioration:

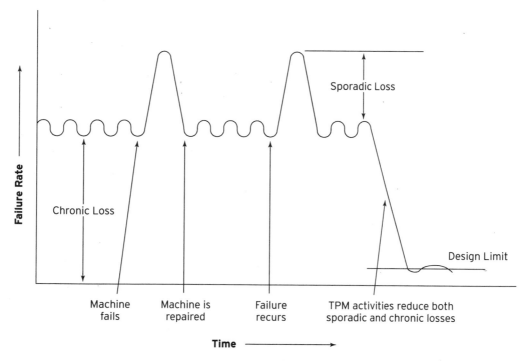

Figure 2-7. Types of equipment losses

- **Natural deterioration:** The deterioration rate expected by the part's designer when used as specified. A component that deteriorates naturally achieves a *natural* or *inherent* life expectancy.
- **Accelerated (or forced) deterioration:** The deterioration rate of a part that is much higher than was expected by the part's designer. Accelerated deterioration is usually caused by the part's being used in an environment where its specified conditions-of-use are not met. A part experiencing accelerated deterioration will have an unnaturally short lifetime.

HOW TPM ATTACKS MACHINE LOSS

We are very aware of machine failures that occur but generally ignore the "seeds" that cause these failures. These unappreciated seeds are minor machine defects. They are generally so slight they are often ignored or overlooked in the belief that they won't cause any problems. However, minor defects, if allowed to exist, will continually and randomly interact in many new ways to create different kinds of failures. In order to prevent machine failure, all minor equipment defects must be detected before they cause failure and then prevented from returning. TPM activities include the relentless pursuit of detecting and correcting all minor machine defects. An iceberg analogy is often used to illustrate this idea, as 90 percent of an iceberg is underwater and hidden from view (Figure 2-8).

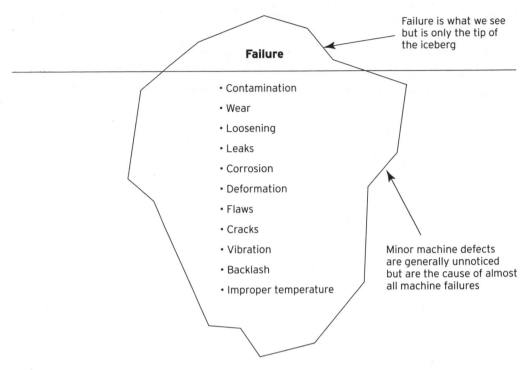

Figure 2-8. Machine failures have many hidden causes

THE TPM PYRAMID OF CHRONIC CONDITIONS

Another way to think about how minor defects cause machine failures is to look at the TPM Pyramid of Chronic Conditions. A pyramid is larger at its base than at its top; the top is supported by the base, which is the foundation of the pyramid. Conditions at the base of the pyramid are numerous and common. Events at the top of the pyramid occur much less frequently. But the events at the top of the pyramid can only occur if the conditions at the bottom exist. Eliminating the conditions that are the root cause of machine failures also eliminates those machine failures (see Figures 2-9 through 2-11).

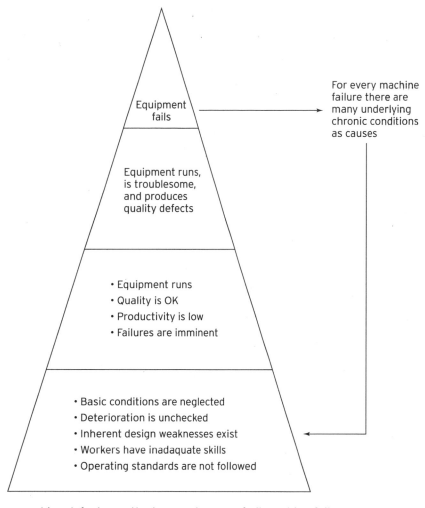

Equipment
fails

Equipment runs,
is troublesome,
and produces
quality defects

- Equipment runs
- Quality is OK
- Productivity is low
- Failures are imminent

- Basic conditions are neglected
- Deterioration is unchecked
- Inherent design weaknesses exist
- Workers have inadaquate skills
- Operating standards are not followed

For every machine
failure there are
many underlying
chronic conditions
as causes

Figure 2-9. Minor machine defects are the true root cause of all machine failures.

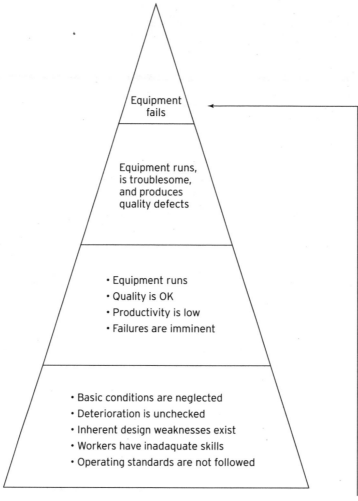

Equipment
fails

Equipment runs,
is troublesome,
and produces
quality defects

• Equipment runs
• Quality is OK
• Productivity is low
• Failures are imminent

• Basic conditions are neglected
• Deterioration is unchecked
• Inherent design weaknesses exist
• Workers have inadaquate skills
• Operating standards are not followed

Most organizations react here and only restore equipment to the point of getting the machine running again. However, permanent improvement is only attained when conditions at the base of the pyramid which cause the failure are improved. Remaining chronic conditions will simply combine in new ways to cause new machine failures.

By not attacking the root cause of equipment failures, reactive organizations are doomed to a cycle of permanent breakdown maintenance.

Figure 2-10. Continually reacting to equipment failures by restoring machine operation does nothing to eliminate the root cause of machine failures

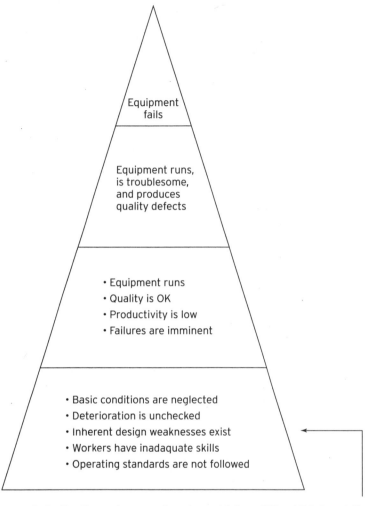

Equipment
fails

Equipment runs,
is troublesome,
and produces
quality defects

• Equipment runs
• Quality is OK
• Productivity is low
• Failures are imminent

• Basic conditions are neglected
• Deterioration is unchecked
• Inherent design weaknesses exist
• Workers have inadaquate skills
• Operating standards are not followed

TPM activities focus on elimination the root causes of equipment failure. Without this foundation of machine weaknesses, equipment failure will not occur.

Figure 2-11. Without a foundation of minor defects, equipment failures cannot occur

TPM FOCUSES ON MACHINE COMPONENT MAINTENANCE

TPM activities focus on machine components and the way that people care for and use them. After all, even large, complex machines are nothing but collections of numerous machine components arranged in a unique way to achieve some desired machine capability. The TPM focus on machine components is the opposite of a traditional product development sequence.

When a new product is first developed, the components that end up on the factory floor in machines to manufacture the product emerge from the following sequence of events:

To improve the product, TPM activities follow a nearly opposite sequence, focusing on equipment components first:

Our commitment to maintaining equipment has already been stated. We believe that:

- Well-maintained machines make many products.
- Poorly maintained machines make fewer products.
- Broken-down machines make none.

This TPM philosophy is easy to state but hard to live up to when our production floor is under fire. Unfortunately, some production managers and maintenance professionals actually believe that more product will be made if machines produce continuously and are only taken down for maintenance when they are broken. We call this "Breakdown Maintenance": run equipment until it breaks, fix it quickly, and then run it until it breaks again.

It seems to go against common sense to take a good machine out of production service to perform scheduled maintenance. However, to achieve high productivity, this is exactly what must be done, because highly productive machines rarely break down. This failure-free state can only be achieved by a rigorous schedule of preventive maintenance.

Maintaining equipment in its optimal state and continually improving its productivity is the whole strategy behind TPM. If we only fix breakdowns, the machine will soon break down again. We must put in place a system that maintains the desired machine state.

Scheduled maintenance must be given high priority in production routines. Carrying out equipment maintenance plans and continually improving those plans is critical. Managers who do not put this principle into practice will never be able to see a TPM program through to success. Instead, they will quickly revert to breakdown maintenance when the going gets tough.

If reactive maintenance is more than 40 percent of your maintenance department's activities, you are not in the maintenance business—you are in the machine repair business.

3

The Team Approach

How do we approach the task of improving
machine productivity?

3

ACTION TEAMS

TPM activities are best accomplished in small groups. At Agilent, these small groups are known as action teams. Members of these action teams are most often selected from different departments—for example production, maintenance, and engineering.

While many problems can be resolved by a single individual or a small group within a single department, other significant problems on the factory floor cannot. They can only be resolved by cross-departmental teams, where each department contributes its own knowledge and capabilities to a coordinated solution. This is one of the principal ideas behind all TPM improvements—shared ownership of a problem among different departments.

Most organizations parse out such problems to a single department. For instance, a product quality problem may belong to the quality department or to the engineering department. An equipment failure would almost certainly belong solely to the maintenance department. This type of "single-ownership system" pits departments against one another as they pursue their own local goals instead of supporting cooperation among them on resolving factory floor loss issues.

 The "Total" in "Total Productive Maintenance" refers to a maintenance strategy that involves the entire organization. Achieving our vision is not a job that the maintenance department can do by itself. It takes work and cooperation from production, engineering, and management as well.

This means we all have to expand the horizons of our current job descriptions. Operators have to help inspect and maintain equipment in ways that are suitable for them. Maintenance technicians must move away from machine repair to activities that actually prevent failure. Engineers must help control machine variability, aid in quality maintenance development, and design processes and products with manufacturing productivity in mind. All of these changes require employees to gain and apply higher level skills than they currently utilize.

Maintenance technicians, of course, will go through a major shift in their daily routines. As we approach our vision of preventing machine failures, technicians will be getting out of the machine repair business and getting into the business of maintaining and improving properly operating machines.

THE TPM PILOT TEAM

TPM is best begun by a pilot team. This is a single team working on, perhaps, a single machine and completing all TPM activities on it to improve its performance. Several things are accomplished by the pilot team:

1. The team learns how to take academic TPM ideas and turn them into real results. TPM is "learning by doing"—no amount of reading or watching will provide the necessary capabilities to complete TPM activities successfully.
2. TPM activities are customized for the organization in which they are going to be used. No two companies are alike. They may use the same machines, but probably use them somewhat differently. Their infrastructures differ, so they do things in different ways. The pilot team has to develop the right way to get things done in its own organization.
3. The pilot team proves that TPM activities can be implemented successfully on equipment in its own factory, using people in its organization.
4. The pilot team develops and provides to the rest of the organization the tools, procedures, and infrastructure to carry out future TPM activities much more easily than the pilot team was able to.

A pilot machine should be chosen with great care. The factory bottleneck machine is not always the best choice because any time that it is kept out of production, factory output is likely lost. A bottleneck machine is a suitable pilot tool if the machine is not used nights or weekends and the pilot team works on it only during these hours. In this case, bottleneck machine improvements are likely to have a significant impact on factory metrics.

In any other situation, however, the best pilot machine might be an unreliable non-bottleneck machine that often starves the factory bottleneck. This machine has some excess capacity, so a pilot team can have it down for some time every week as they learn to apply TPM steps to the machine.

A complex and troublesome machine is also a good candidate for a pilot machine. People tend to believe if TPM methods improve this type of machine, they can improve any other machine on the factory floor.

In Agilent's pilot program, a $3 million cluster tool that deposits oxide on wafers was used. Agilent owns nine of these machines and they are very nearly—but not quite—the production limiter in the factory. The machines often broke down and had many quality control problems. Sometimes problems would keep some of these machines down for weeks while technicians puzzled over how to repair them. Without productivity improvements, more of these machines would have to be purchased. However, with limited clean-room space, buying more machines was not a realistic option. The best solution was, obviously, to get more output through the machines that we already owned.

The pilot team was able to reduce equipment failures by 90 percent on the pilot machine and improve its throughput capability by 50 percent in less than a year without a single design change to the machine. The maintenance work required to keep the machine in its new state was also less than the amount of repair work formerly needed to keep the machine running—not more work, as most people had expected.

At the outset of TPM, a pilot team must master the steps, prove they improve machine performance, and prove they make work easier for those who operate and maintain the machines. This can greatly reduce the amount of organizational resistance to change that stifles many TPM activities.

Once the pilot team has completed its work, more teams can begin to follow in their footsteps. At Agilent, we followed the pilot team with five more teams on five new types of equipment. These teams, which included all of our maintenance and production supervisors as team members, were referred to as Manager's Model Teams. The challenge to each team was to learn and then apply the techniques created by the pilot team to their own machines. Each team, of course, faced a slightly different challenge in accomplishing this.

At this point, the biggest challenge is to transfer new knowledge horizontally throughout the organization. This process must begin with managers and may be easy or difficult, depending on the size and complexity of an organization. Manager's Model Teams not only improve the performance of more equipment, but also develop knowledgeable leadership for the TPM equipment teams that follow.

By the time the Manager's Model Teams have succeeded in their efforts, the organization will have created many capable teachers of the TPM steps, and the number of teams can then grow exponentially. This idea is illustrated in Figure 3-1. This appears to be a slow way to implement TPM, but many companies that have skipped the pilot and model team phases and proceeded directly to large-scale implementation across all of their departments have met with such strong resistance that they could not continue successfully.

Agilent's experience is that the year taken by the pilot team to prove TPM success and develop the support structures for new behaviors was critical to further the progress of TPM activities on other equipment. Rushed preparation leads to poor TPM deployment.

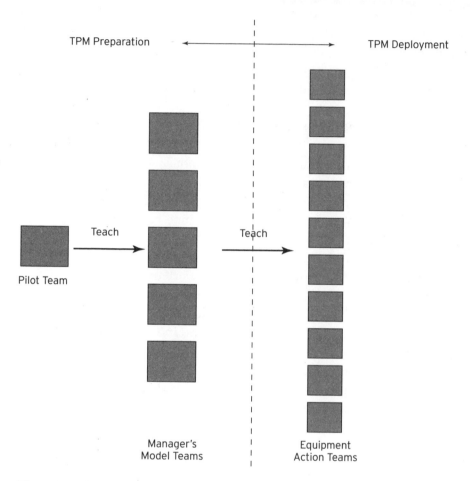

Figure 3-1. The deployment of TPM action teams

IMPLEMENTING TPM ACTION TEAMS

The goal of each TPM action team is to get out of the business of repairing broken machines and into the business of preventing machine failures and other productivity losses. To accomplish these goals, each team implements a systematic machine improvement method. The following is a generalized approach to TPM team implementation:

1. Deploy cross-departmental teams, ideally involving everyone in the organization, to all machines on the factory floor, one team per type of machine.
2. Train each team member to competently use TPM tools and productivity improvement methods.
3. Charter each team to attack the productivity losses on their equipment—first preventing machine breakdowns, and then reducing other types of losses.
4. Challenge these teams to achieve a rapid improvement rate, typically a minimum of 70 percent improvement within a year. Specific machine losses should be reduced to zero.
5. Deliver contingent consequences to team members for their contribution to successful results: greatly reward the highest level contributors, and help others to improve their level of contribution.

TPM team members should work in the same area at the same time as much as possible, so they can work together effectively to advance TPM activities.

TPM teams should also include employees from various levels in the factory—managers, operators, maintenance technicians, and engineers. Night-shift teams at Agilent often meet in the early evening or morning hours to help accommodate engineering and management participation.

Ideally, every person in an organization should be an active member of a TPM team. However, this was very difficult for Agilent to achieve because people on four different shifts perform the same work on the same equipment but have very little contact with each other. So Agilent created TPM teams that consist of both core members and associate members. Core members work on the same shift. They meet regularly and are chartered to use the TPM steps to drive improvements. The core members then transfer those improvements to all the associate members. Associate members support TPM activities by providing core members with their machine issues, observations, data, and ideas for improvements.

Of course, Agilent operators and technicians work on numerous machines, so they may be associate members on one machine team and core members on another team. Ideally, every person is a core member on one TPM team.

Figure 3-2 shows the relationship between the TPM team core members and associate members at Agilent.

Many companies wonder about the wisdom of including machine operators in TPM maintenance activities. However, it is foolish to exclude them because they spend more time with the equipment than anyone else; they can detect many minor equipment abnormalities long before they are seen by maintenance technicians or engineers. Machine operators can also perform certain minor maintenance routines more efficiently than maintenance techs.

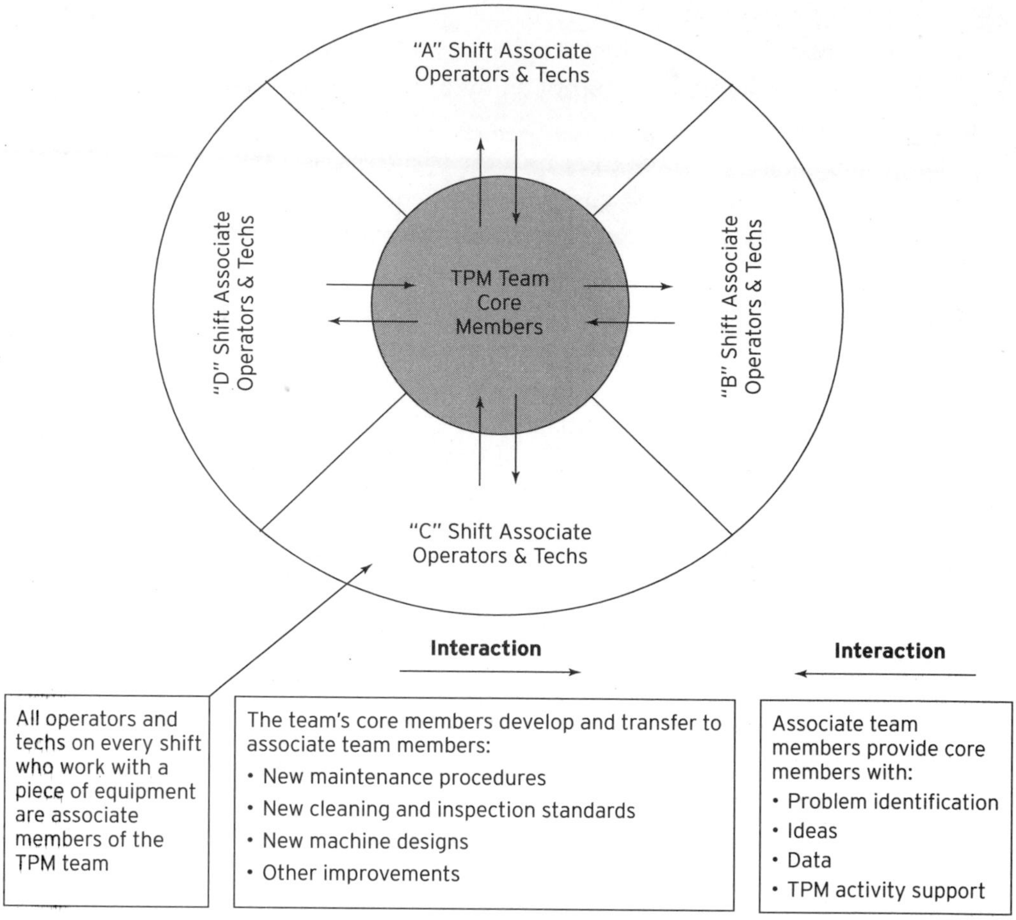
"A" Shift Associate Operators & Techs
"B" Shift Associate Operators & Techs
"C" Shift Associate Operators & Techs
"D" Shift Associate Operators & Techs
TPM Team Core Members
Interaction
Interaction
All operators and techs on every shift who work with a piece of equipment are associate members of the TPM team
The team's core members develop and transfer to associate team members:
• New maintenance procedures
• New cleaning and inspection standards
• New machine designs
• Other improvements
Associate team members provide core members with:
• Problem identification
• Ideas
• Data
• TPM activity support

Figure 3-2. The TPM team concept with core and associate members

KEYS TO TPM SUCCESS

There are two basic requirements for TPM success.

1. Know the TPM steps, and follow them. It wastes time and other resources to try to re-invent TPM. It is important to learn from others who have succeeded at creating a factory with world-class levels of productivity. TPM cannot be learned simply by reading a book or going to a lecture.

   Improvement teams that do not understand the steps of TPM tend to fall back on the usual approach to machine improvement—make a Pareto chart of the machine's failures and, starting with the largest one, beat it down by any and all means possible. A team of people can almost always do this successfully, but the machine will simply produce new kinds of failures faster than the team can fix them. This is not the TPM approach to failure prevention.

2. A company must have management commitment and competency to lead the change. There is a big difference between managing the day-to-day operation of a factory and leading people through changes in the nature of their work.

   TPM machine improvement teams must be guided through the improvement steps on their machines. These steps will require significant and permanent behavior changes on their part. Without knowledgeable guidance from their leaders, these changes will not occur and the old way of operating and maintaining equipment will prevail.

   Everyone in the organization must participate in TPM activities. A company cannot assign some people to be involved in TPM and not others—TPM is a factory operating system change program that affects the way that all employees perform their routine jobs.

   Employees won't change how they work because a TPM consultant—whether from outside the company or the TPM program office manager—wants them to. Employees change only when their own managers want them to and reinforce the changed behavior properly.

PART 2

Step-by-Step
TPM Implementation

A TPM action team implements the following five steps to
improve equipment performance.

1. **Restore equipment to "new" condition:**
   - All areas of the equipment are clean and free of humanly detectable minor defects
   - Cleaning and inspection standards are created to keep the machines in this condition
   - Machines and their work areas are visually controlled as much as possible

2. **Identify complete maintenance plans:**
   - PM checklists
   - PM schedules
   - PM procedures
   - Inspection specifications
   - Replacement part numbers
   - Part logs
   - Quality checks

3. **Implement maintenance plans with precision:**
   - Complete PMs on time
   - Complete PMs 100 percent—no checklist items are skipped
   - Execute PMs without variation, no matter who carries out the PM
   - Continually advance the knowledge and skill of factory floor personnel

4. **Prevent recurring machine failures:**
   - Implement failure analysis to prevent recurring failure
   - Implement continuous PM evaluations and improvement—"easier, faster, and better"

5. **Improve machine productivity with the following methods:**
   - Lubrication analysis
   - Calibration and adjustment analysis
   - Quality maintenance analysis
   - Machine part analysis
   - Condition-of-use and life analysis
   - Productivity analysis:
     – Availability
     – Setup
     – Scrap
     –Yield
     – Minor stoppages
     – Speed
   - Extended condition monitoring
   - Continuous condition monitoring
   - Maintenance cost analysis

These TPM steps are not completed serially, as the step numbers seem to imply. To maximize progress, some of these steps should be performed in parallel.

The table that follows shows a typical TPM master plan and the critical path of TPM improvement activities.

Typical Five-Step TPM Master Plan

Milestones	Improvement Activities	Support Activities	Typical Time Line
	I) Eliminate MInor Defects	**II) Identify Maintanance Plans**	
		III) Develop Precision Maintenance	
Equipment is clean and free of minor defects			One Year
	IV) Prevent Recurring Failures		
Failure prevention is a normal part of maintenance work			Two Years
	V) Reduce Productivity Losses		
			Three Years
			Four Years
A new organizational culture is created	Continual improvement to zero loss is part of everybody's daily working routine	Continuous learning is a normal part of everybody's daily working routine	World-class factory productivity is achieved

4

Step 1: Restore Equipment to "New" Condition

This chapter describes how to implement TPM Step 1, including:

- Setting specific goals for equipment, operators, and technicians
- Detecting minor machine defects
- Setting component standards
- Developing Step 1 support tools
- Getting TPM teams started
- Working safely
- Advancing Step 1 activities
- Sustaining Step 1 gains

STEP 1 GOALS

In Step 1, the following activities are performed to restore equipment to "new" condition:

- All areas of the equipment are clean and free of humanly detectable minor defects
- Cleaning and inspection standards are created to keep the machines in this condition
- Machines and their work areas are visually controlled as much as possible

To determine whether they have met the requirements for Step 1, teams examine whether they have completed the following measurable goals. (A detailed audit form is provided at the end of this chapter for teams and managers to use to determine whether they have met these goals.)

- **Goals for equipment.** The following activities help to create stable equipment conditions, stop accelerated deterioration, and improve machine predictability:
  - Restore a machine to its clean and defect-free condition
  - Create a cleaning and inspection standard to keep it that way
  - Eliminate sources of contamination and improve maintenance access

- **Goals for operators:**
  - Learn new skills to assist technicians in basic maintenance
  - Carry out the cleaning and inspection standards as part of production's daily routine
  - Learn that minor defects cause many machine failures
  - Develop a "good eye" for detecting minor equipment defects
- **Goals for technicians:**
  - Develop a "good eye" for detecting minor equipment defects
  - Learn that minor defects cause many machine failures
  - Maintain machine components "as they should be"

MINOR MACHINE DEFECTS

Types of Machine Defects

There are three categories of defects:

- *Major defects* shut the machine down and prevent it from running altogether. Usually a broken machine will contain only one major defect that can be corrected with nominal troubleshooting and repair skills.
- *Medium defects* impair the machine's performance but usually won't stop it from running. It may run more slowly or produce product with a lower quality level. Often, machines exhibiting this behavior have several medium-level defects. Also, problems are often intermittent.
- *Minor defects* by themselves seem to do no harm. This is probably because we can observe a machine with several minor defects and notice that it appears to be running just fine. Minor defects don't seem to matter at that moment. But minor defects, left unattended, hurt machine performance in many ways. Most machines have dozens of minor defects at any one time; in fact, Agilent found that many of our machines contained over 100 minor defects.

  Eliminating minor equipment abnormalities seems contrary to common sense. It would seem more practical to work on major equipment problems rather than minor defects; unfortunately, this assumes that minor defects don't matter and play little role in machine failures. This is a false belief: minor defects, taken together, are often the primary cause of machine failures.

Component Standards: How Components "Ought to Be"

Any component in the machine that is not the way it should be is a minor defect. In order to identify machine abnormalities, team members must know how components "ought to be." We call this the *component standard*. In this first TPM step, we restore all humanly detectable minor abnormalities in the equipment to the standard that we have set for them.

Minor defects are often caused by normal machine deterioration. This principle is illustrated in Figure 4-1.

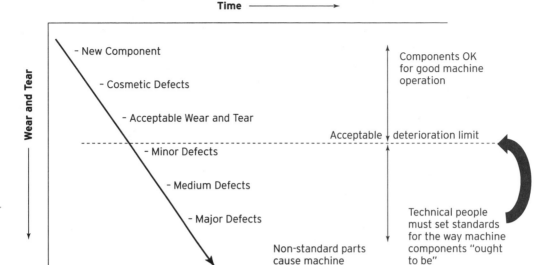

Figure 4-1. Normal machine component deterioration

However, machines also contain many abnormalities that are not caused by deterioration. In Agilent's experience, about half of all the abnormalities found in our equipment were man-made, caused either by something we did or something we didn't do. We overtightened a bolt, left out a flat washer, used the wrong lubricant on a ball bushing, or lubricated a plastic bushing that was supposed to remain dry. We made adjustments improperly. These and many other man-made problems occurred in Agilent's equipment despite a highly trained technician workforce. In each case, technicians thought they were doing the right thing, but lack of detailed knowledge caused their well-intentioned efforts to go astray.

Figures 4-2 and 4-3 show a typical man-made defect and its solution. The hose installed on this hot water circulator is the wrong type of hose. It does not have the temperature rating required for the water temperature in use and does not have the proper inside diameter to fit properly on the equipment's barbed hose fittings. These defects caused the hose joints to leak; hose clamps then were added in an attempt to control the leaking. In this case, the wrong hose was used simply because the technician who installed it did not know this equipment required another type of hose.

 The lack of knowledge of a thousand seemingly insignificant details contributes in a huge way to many equipment losses.

Figure 4-2. A man-made equipment defect—the wrong hose is used on these barbed fittings. Hose clamps, which are not required on this type of barbed fitting, have been added to try to control leaking at this joint.

Figure 4-3. This equipment has the correct hose installed on its barbed fittings and will not leak at the pressure and temperature of the hot water in use, even without hose clamps

About half of all equipment abnormalities are man-made and **not** *caused by machine deterioration, as most people suppose!*

Operator behavior also has caused failures in Agilent equipment. A machine with a powerful heat lamp was having a high rate of lamp failures. One of the root causes contributing to the failure rate was that operators opened the lid to load wafers and then dropped the lid the last inch when closing it. Even a one-inch drop of this heavy lid sent shock waves through the machine and the lamp. This vibration reduced the lamp filament life. Once the operators learned about the problem they were causing, they were careful to set the lid down gently.

Cleaning to Detect Machine Defects

The first activity of a TPM action team is to use their human senses to detect abnormalities in the equipment. This begins with machine cleaning.

Cleaning serves two purposes:
- First, it removes contamination—which is usually considered a minor defect—because contamination often causes high rates of machine wear and can also cause quality problems.
- Second, cleaning is a very good way to inspect machine components because the cleaner's eye and hand dwell on a machine component for a sufficient length of time to really "see" how it is. Casual visual inspections are often not adequate to detect minor problems.

  Cleaning some of Agilent's clean-room equipment is impractical because the environment keeps the machines clean beyond the ability of a human to sense contamination. In these cases, inspections are often performed by looking at and touching the parts. Touching improves the ability to sense minor abnormalities.

Eliminating minor defects in equipment is not a new idea, even though it may seem unconventional to many maintenance technicians. The author's high school automotive teacher taught this principle to improve a car's engine performance: "Before you reach for your troubleshooting manuals and oscilloscopes, or before you spend any serious effort looking for solutions to complex problems, open the hood and look for the obvious. Is there corrosion on the battery terminals? Are wires hanging loose or showing frayed insulation? Is there any type of fluid leaking? Is the carburetor or choke dirty? Are the sparkplug wires fully seated? Is the air cleaner dirty? Are any bolts loose or missing?" Before doing any higher technical work, he recommended keeping these and other humanly detectable conditions "the way that they ought to be."

> *Most minor defects are detectable by our human senses. We see them, hear them, smell them, and feel them. (We do not use "the taste test" in our factory.) With experience, people performing these inspections develop a "good eye" for the types of minor defects that can occur in their equipment and learn to spot them with ease.*

Examples of minor defects that can be detected by human senses include:

- Dirt or contamination
- Liquid or air leaks
- Loose or missing bolts and fasteners
- Gauges not reading in their "standard" range
- Worn wiring insulation
- Wear in plastic tubing or similar parts
- Loose or deteriorated connectors
- Machine motions that are not smooth or accurate
- Switches that are not set in the correct positions
- Sensors that are not aligned or working correctly
- Fluid reservoirs that are not filled to the correct level
- Locking pins and devices that are missing or loose
- Safety guards that are not in place
- Surfaces showing signs of wear (detectable by the repeated appearance of contamination)
- Machines that sound abnormal
- Machines with unfamiliar odors
- Machines with unfamiliar vibrations

Minor defects are classified as functional or cosmetic. TPM is not concerned with cosmetic defects. Keeping a machine clean and neat enhances a person's ability to inspect the machine and discover minor defects. For example, if the bottom of the machine is full of collected material such as nuts, bolts, and other miscellaneous materials, then the human eye will not be able to detect a loose bolt that has recently fallen into this area from the machine above.

Likewise, in a very dirty area no one would likely notice a new spot of contamination being formed by a fraying belt. But suppose the machine floor is kept pristine clean. The discovery of a bolt lying in this area now has meaning. We will look upward in that area to try to discover the source of the lost bolt. The same would be true for a new collection of debris. This is one of the reasons for our goal of being clean and defect free: it enhances a person's ability to detect minor defects.

"Open the hood and look for the obvious." *Make these easy-to-spot defects "as they should be" before attempting any higher-level equipment improvement efforts. This is what Step 1 of TPM is all about.*

Minor Defects Cause Machine Failures

The first significant change in thinking that team members experience is realizing the importance of correcting minor machine defects. In the past, they were probably only concerned with defects they thought "mattered." Most minor defects, of course, were seen as insignificant. In Step 1, team members begin to understand their importance. Minor defects cause machine failures in at least four different ways.

1. Minor defects hurt machine productivity by continuing to deteriorate and growing into medium and major defects. After all, every medium and major defect was once a minor defect or resulted from minor defects. Left unattended, minor defects only become worse. It is better to nip them in the bud instead of letting them go until larger, more difficult, more expensive repairs are required.

2. Minor defects impair equipment performance by causing accelerated deterioration in other parts. Even though a machine is running fine today with minor defects, these defects often cause a faster rate of deterioration in surrounding parts, producing more frequent machine breakdowns.

   For example, the loss of one small wheel weight on an automobile tire is a minor defect. It may cause only a very slight vibration, and that only at a certain road speed. But this minor vibration induces a higher rate of wear in many expensive front-end parts such as wheel bearings, struts, rod ends, and tires, deteriorating them more quickly than normal. Higher rates of wear cause higher failure rates for the machine, more maintenance work, and higher operating costs.

3. Minor defects hurt machine performance by interacting with one another while they are all still minor in random ways that, together, cause machine failures. These kinds of problems are notoriously difficult for technicians to repair because they most often happen intermittently—only when certain minor defects interact in just the right way. Typical troubleshooting methods do not diagnose this type of failure because there appears to be no repairable problem with the machine.

   It is often assumed that these problems are caused by design limitations in the machine. For example, a machine handling IC wafers may occasionally break a wafer and then run fine for days with no further trouble. Then the machine will break another wafer. This type of intermittent problem could easily occur because of numerous interacting minor equipment defects.

4. A single minor machine defect may, all by itself, cause a variety of machine failures to occur without any interaction with other minor defects. This is not a frequent occurrence, and not every minor defect will cause equipment failure or productivity loss all by itself. Still, no one can tell which minor defects will cause such performance problems; therefore, all minor defects must be eliminated.

Our work on many pieces of equipment in Agilent's IC fab has proven that eliminating humanly detectable minor defects in equipment can significantly reduce the equipment failure rate, even though no single defect seems important by itself. We achieved a failure rate reduction of 40 percent in our IC fab by eliminating a number of minor defects. The results of such efforts are shown in Figure 4-4.

The most common kinds of defects corrected were: loose and missing bolts; electrical wires with cracked or frayed insulation; pneumatic and liquid leaks; kinked or worn hoses; contamination; improper adjustments; and gauges that indicated abnormal conditions. These minor defects were easy to spot and, in almost all cases, easy to correct.

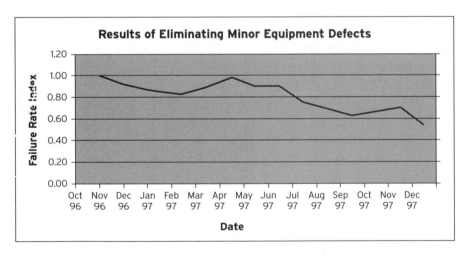

Figure 4-4. A 40 percent reduction in all factory machine failure rates was achieved simply by eliminating obvious minor defects in the machines

Examples of Minor Equipment Defects

Figures 4-5 through 4-11 show a variety of minor equipment defects found in Agilent's IC fab.

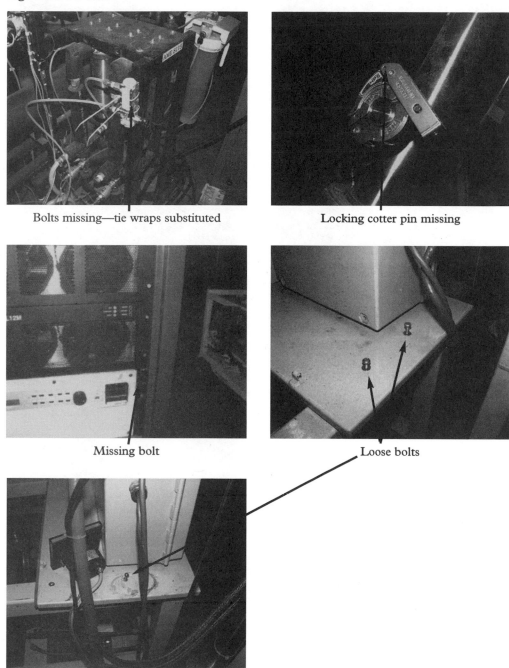

Bolts missing—tie wraps substituted

Locking cotter pin missing

Missing bolt

Loose bolts

Figure 4-5. Minor defects associated with bolts and fasteners

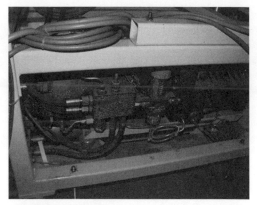

High contamination levels are minor abnormalities

Before the bottom of this machine was kept clean, the sudden appearance of contamination from the drive belt above this area would have gone unnoticed.

Figure 4-6. Minor defects associated with contamination

Gauges indicating out-of-spec conditions

A 160-psi gauge that is very difficult to
check for proper condition

Gauges with no indication of
what they should read

Figure 4-7. Minor defects associated with machine gauges and visual controls

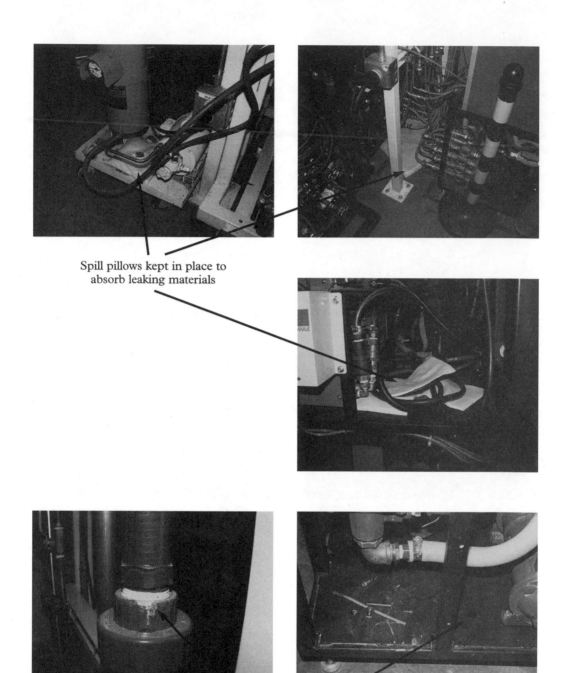

Spill pillows kept in place to
absorb leaking materials

Slow leaks

Figure 4-8. Minor defects associated with leaking equipment

A flexible connector misused—
flexed beyond its capabilities

A loose filter-mounting device

This compressor is very noisy
(detected audibly) and has a severe vibration
(detected by touch)

Figure 4-9. Other humanly detectable minor equipment defects

Is this part missing from its place in the machine, or was it just left here?

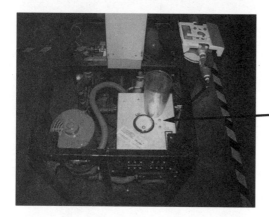

Numerous materials that do not belong on this machine

A mislabeled machine control system. What machine does it control?

Figure 4-10. Minor defects associated with clutter and equipment labels

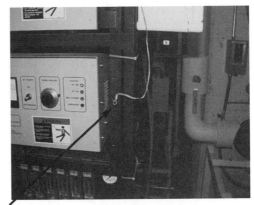

Unattached ground lugs

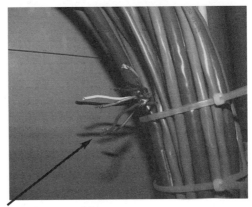

Exposed wire ends that connect to nothing

Figure 4-11. Minor defects associated with electrical wiring

Using Appropriate Team Members to Set Machine Standards

Because a TPM team is cross-functional, it should always be possible for some member of the team to define how any machine condition "ought to be"—that is, define the component standard.

Not all standards should come from engineers or managers. On the contrary, component standards are best set by the appropriate team members. For example, production-line standards should be set by the people who work most closely with the condition in question, as long as satisfactory results can be obtained. Figure 4-12 illustrates this idea. (However, be cautious in the early stages of TPM. People tend to include minor defects in component standards as the way a part "ought to be." TPM activity leaders must prevent this by helping people truly understand minor defects.)

Agilent's production operators are continually undertaking higher-level maintenance tasks. Table 4-1 is an example of how a TPM team might sort out maintenance chores between operators and techs on an automobile in the early stages of team development. Of course, at a later time, these roles could and should be elevated to ever-higher levels.

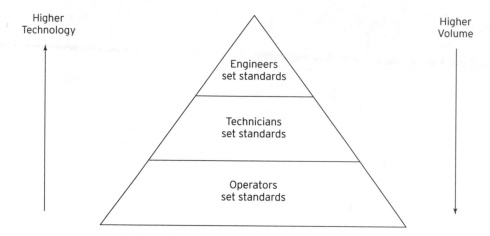

Figure 4-12. Machine standards need to be set by the people in the best position to know what they should be

Table 4-1

Work Performed By	Routine Maintenance (Repeated Frequently)	Time- or Use-Based Maintenance	Condition-Based Maintenance

• The car owner and operator	• Keep the car clean
	• Check fluid levels
	  – Fuel
	  – Coolant
	  – Washer fluid
	  – Power steering
	  – Auto transmission
	  – Engine oil
	  – Brake fluid
	• Keep tires inflated
	• Inspect for minor problems:
	  – Handling
	  – Noise
	  – Vibration
	  – Leaks
	  – Dash gauges
	  – Battery corrosion, etc.
	• Comply with conditions-of-use
		• Replace PCV valve
		• Replace engine coolant
		• Replace fuel filter
		• Check parking brake operation
			• Replace wiper blades
			• Replace air cleaner

• An automobile maintenance technician	• Change engine oil
	• Lubricate chassis
	• Check fluids:
	  – Differential
	  – Manual transmission
	• Clean and inspect chassis components
		• Replace spark plugs
		• Adjust auto transmission bands
		• Change transmission fluid
		• Align front end
		• Rotate tires
		• Repack wheel bearings
			• Inspect/service:
			  – Belts
			  – Hoses
			  – Brakes lines
			  – Brake pads
			  – Front-end components
			  – Muffler and tailpipe

STEP 1 TOOLKIT

Every team working to eliminate minor abnormalities in its equipment needs a set of appropriate tools, which must be developed and made available to every member. The tools described in this section include:

- TPM program safety
- The "Three lists"
- Cleaning and inspection materials
- M-tags
- 1-point lessons
- Visual controls
- Team notebooks and activity boards
- Cleaning and inspection (C&I) standards

TPM Program Safety

No equipment improvement plan that involves people in changing roles can afford to overlook the safety risks of such activities. As people engage in new activities, the safety of these new activities must be assured.

In carrying out all TPM activities, Agilent always provides for safety first. While Agilent's manufacturing safety assurance program began long before its TPM activities, safety assurance is an important part of continuous improvement, and safety management must be incorporated into Step 1 activities.

Figure 4-13 is an overview of Agilent's safety program.

Figure 4-13. Agilent's safety program

Three key factors work together to assure safety for our employees:
1. Safety management systems.
2. A safe working environment.
3. Safe employee behaviors.

These parameters are built into every manufacturing process and procedure and are all continually being improved.

Safety Management Systems

Safety management requires that safety considerations be built into every aspect of the IC fab operation. Safety is not a project or extracurricular activity, nor are safety improvements reactive—making changes only after someone gets hurt. Safety is part of every aspect of our day-to-day work routine.

Every written procedure, whether it is used by operators, technicians, or engineers, includes a section on safety. Only people certified to do the procedure are allowed to carry it out, and certification always includes an understanding of the safety requirements of the procedure. Every employee's evaluation includes a rating on safety performance—not only behaving safely, but also contributing to safety improvement. Employee safety is our number one business priority, and it is always listed as such on our business objectives.

Every safety incident, whether or not it involves an actual injury, is investigated thoroughly, and corrective action is taken. Safety incident reviews are the top priority for all management personnel. Our philosophy is simple: "No one gets hurt—not even a scratch."

A Safe Working Environment

Our second safety factor is a safe working environment. An IC fab has many safety hazards—numerous machines, many hazardous chemicals, and a very complex routine of human activity taking place in a tightly packed clean-room environment. As much as possible, the IC fab is engineered to be an inherently safe environment. All OSHA requirements on safe design are followed in detail; Agilent often sets more stringent standards than those legally required. Chemical systems, for example, have many redundant levels of safety built into their operation; even multiple equipment failures or human errors will not produce an injury.

The IC fab safety design is also continually being scrutinized and improved. A standing safety committee is dedicated to continually improving the safety design of our workplace. Monthly safety audits of the facility are always conducted and ways are sought to improve the safe design of the workplace.

Surprisingly, the most common type of injury in the IC industry is not related to mechanical or chemical hazards. Instead, it is repetitive stress injury—a condition that arises from stressful ergonomic human motion, repeated many times over many years. A common ergonomic problem involved a simple wrist rotation that was necessary to get a cassette of wafers out of a box and into a machine loading station, as shown in Figures 4-14 and 4-15. To eliminate the wrist rotation and prevent injuries, "ergo loaders" were installed on every machine loading station, as shown in Figures 4-16 and 4-17.

A daily stretching program was also put in place for operators to help reduce ergonomic impacts from their repetitive job routines.

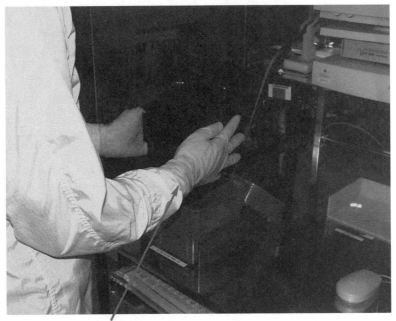

Figure 4-14. A neutral position of the wrists is naturally maintained when removing a wafer cassette from a transport box

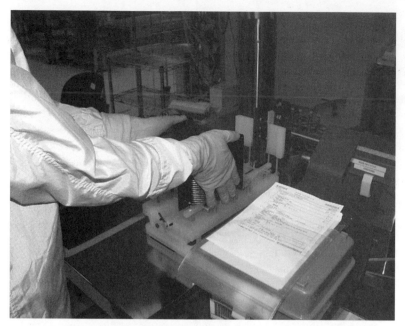

Figure 4-15. The wrists must be rotated to set the wafer cassette in a vertical position in this old wafer transfer device. This rotation, repeated long enough, can cause damage to the operators' wrists.

Figure 4-16. Adding this "ergo loader" to the machine allows the operator to maintain a neutral wrist position for the entire loading operation

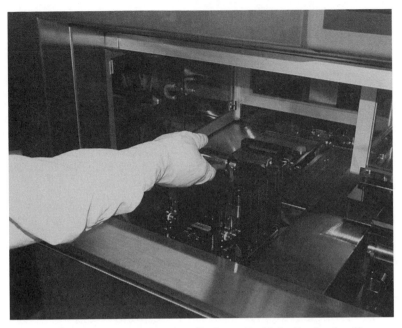

Figure 4-17. Once the wafer cassette is nestled into the loader, the loader handle is lifted to rotate the cassette 90° to an upright position

Safe Employee Behavior

Safe employee behavior is the third leg of Agilent's safety program. No environment, no matter how well managed or designed, can remain safe if people do not behave safely within it. For example, safety glasses with side shields are required of every occupant of the IC fab at all times. This behavior is taken seriously and is enforced by every person working on the factory floor. People who inadvertently walk into the IC fab without the proper eye protection do not get very far before someone points out to them that they must immediately obtain safety glasses.

Agilent operates a proactive behavioral safety program to continually improve the safe working habits of our IC fab employees. Safe behavior does not emerge solely from instructions to "work safely." This is a noble attitude, but it must be backed up with an improvement process if it is to be realized.

Agilent's program to create safe working behaviors is called BSAF, which stands for Behavior Safety Awareness is Fundamental. The program works along the following lines:

- A behavior safety committee—made up of factory floor employees and managers—creates and updates a checklist of behaviors that have been identified as important for the safety of all IC fab personnel.

- Everyone working in the IC fab is trained as a behavioral safety observer—able to observe others behaving correctly or incorrectly according to the items on the checklist.
  - Basic training is conducted on the checklist of safety behaviors.
  - Training certification is only completed after observation skills are developed through practice—people make five observations of other people's behavior every month for three months before they are certified as safety-behavior observers.
  - Every employee completes at least one behavioral safety observation of fellow employees each month.
- Industry studies have shown that compliance to safe behaviors exceeding 95 percent is required to achieve a significant reduction in the illness and injury rate. Our people strive for 100 percent compliance with the checklist of safe behaviors.

Agilent also maintains a highly trained HAZMAT (hazardous materials) team on all shifts. These people are trained to respond to emergencies—whether involving people or the facility.

Safety is a process like any other. In order for this process to be carried out as designed, safety metrics are defined and measured. The results of these measurements are posted on safety activity boards for all to see. Figures 4-18 and 4-19 show some of these safety activity boards with data from all of the above-listed safety activities posted. Our various safety committees constantly strive to improve results:

- Ergonomic improvement projects are on schedule.
- All safety incidents are reviewed, and corrective action is taken in a timely manner.
- Monthly safety audits of the facility are made on time.
- All deficiencies found during these audits undergo high-priority design improvements.
- All HAZMAT teams are properly staffed and are continually recertified on HAZMAT procedures.
- BSAF program observations, training, and results are progressing satisfactorily toward our behavioral safety goals.
- The daily stretching program receives full participation.

The "Three Lists"

To keep operators on TPM action teams safe as they got more involved in machine maintenance, Agilent created a new safety activity referred to as the "Three Lists": three equipment preparation lists that provide step-by-step instructions for cleaning and inspection. The lists are particularly important

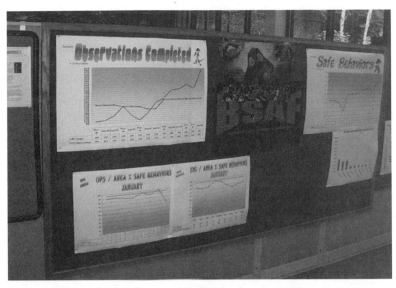

Figure 4-18. The BSAF Behavioral Safety Program Activity Board with observation data and measured results

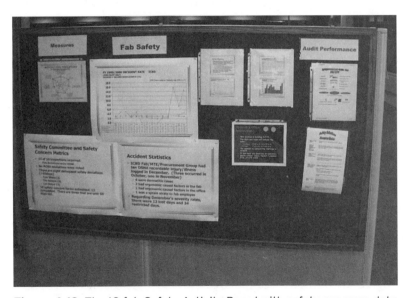

Figure 4-19. The IC fab Safety Activity Board with safety program data and results posted

for production operators participating in a TPM action team who do not have the same level of knowledge about working on equipment safely as do machine technicians.

A management safety team, which usually consists of the TPM coordinator at the site, a maintenance manager, and a representative from the site's safety department, creates lists for each type of machine. Furthermore, this committee is assisted on each machine investigation by a senior technician who services the machine being prepared for TPM activities. Agilent's Three Lists are:

- *List 1: Prepare the machine for safe access and cleaning.* This list is a step-by-step instruction for placing the machine in a safe state so that no automatic machine activity will take place. This may include shutting down power to the machine or placing it in a maintenance or other nonproduction mode. Some machines may require that certain components—such as vacuum-chamber gate valves—be placed in a closed position, or that load-lock doors be placed in an open position. It is up to the safety team to consider all that must be done to prepare the machine for safe team access—especially safe access by operators on the team. It is important to comply with all legal lockout/tagout and other safety requirements when preparing this list.

- *List 2: Identify the accessible and inaccessible areas of the machine for team access.* This list specifically defines all areas that can be safely accessed by all members of the team. Teams generally stay out of high-voltage areas and areas containing hazardous conditions. Areas that should be made safe for operators include all exposed areas and areas behind covers that can be removed without tools. Any area defined as accessible must be made absolutely safe during the cleaning and inspection activities. All hazards must be eliminated, including high-temperature surfaces, high-voltage sources, high-current sources, chemical hazards, pressurized liquid or gas hazards, sharp objects, pinch points, and sources of potential energy, such as falling weights, pressurized pneumatic devices, or charged capacitors.

  Agilent engineers created many guards for hazardous devices that existed inside otherwise large safe areas. This is an extremely important step. Areas cannot be declared accessible to the team with a caveat such as: "Be careful not to touch the high-voltage terminals." An area must be made absolutely safe to be accessible, or it must be declared inaccessible for the team. In such a case, only maintenance technicians with the safety training required to access the area can perform the work required to eliminate minor defects there. These areas must remain inaccessible to the rest of the TPM team unless they can be made safe for everyone.

- *List 3: Prepare the machine for production service.* This list describes in detail every step that must be taken by the team after their TPM activity, assuring that the team's activities have not altered the machine's performance in an unexpected way. It should include picking up all of the tools and cleaning materials used during this activity, returning all removed machine covers to their normal position, and placing the machine controller in its "production ready" mode. It may also include some type of qualification test. The machine's process engineer is best positioned to determine this particular need, based on the team's access level and maintenance activities.

Examples of the Three Lists for two of Agilent's production machines follow.

Machine: Wafer Anneal
I) Preparation of the Wafer Anneal for operator cleaning and inspection: • No preparation needed—proceed with the machine in production-ready mode
II) Accessible and inaccessible areas of the Wafer Anneal: • Accessible: any door that does not require tools to open it • Inaccessible: any door that requires tools to open it
III) Return to Production: • Clean up all tools and cleaning supplies • Close all access panels • Return all surrounding equipment to its proper place

Machine: Metal Dep
I) Preparation of the Metal Dep for operator cleaning and inspection: • Turn off all six ion gauges • Close all six Hi-Vac chamber valves • Place the Metal Dep controller in "Standby" mode • Disable the control panel by placing its key in the "REMOTE" position and removing the key
II) Accessible and inaccessible areas of the Metal Dep: • Accessible: all areas of the machine are safely accessible, except those areas specifically listed as inaccessible • Inaccessible: any door that requires tools to open it or is labeled "High Voltage"
III) Return to Production: • Clean up all tools and cleaning supplies • Close all access panels • Check the interlock panel for three lights—Cathode, H_2O, and Panel—to make sure all interlocks are properly set • Insert the control-panel key and enable the control panel by turning the key to the "LOCAL" position • Return the system to "READY" condition. This will automatically open all of the Hi-Vac valves and turn on all but one of the ion gauges • Turn on the load-lock ion gauge

Figures 4-20 through 4-22 show examples of work that was done to various machines to safely prepare certain areas of equipment to be accessible.

Figure 4-20. A plastic cover was added to protect against high-voltage terminals

The machine in Figure 4-20 originally contained interlock switches on every access door, which dropped power to the entire machine any time that any door was opened. Once the machine was powered down, it often took over an hour to warm it up again to continue processing wafers. This feature was very safe, but it made maintenance access to the machine very difficult. This discouraged PMs and encouraged breakdown maintenance.

A thorough safety investigation revealed only one unsafe point behind this set of double doors—a power supply with 120-volt terminals exposed. By enclosing the terminals inside a clear plastic case, the door interlocks could be removed. Now this area of the machine is listed as "accessible" in the second list because there is no longer any safety hazard behind these doors. The area is now very accessible for cleaning, inspection, and PMs. Also, a major productivity loss—the warm-up time after the door was opened—was eliminated.

The machine in Figure 4-21 also required the addition of some protective guards to make major areas safe and accessible.

Figure 4-21. Guards were added inside a machine to protect fingers from pinch points. These simple plastic guards made this entire area of the machine accessible to the improvement team.

Before a catwalk was constructed for the machine in Figure 4-22, the area underneath was a rat's nest of wires, cables, and hoses. This area could not be inspected effectively, nor was it safe. People entering this area for maintenance purposes were apt to do more harm than good because they had to step on all of these items to access the area.

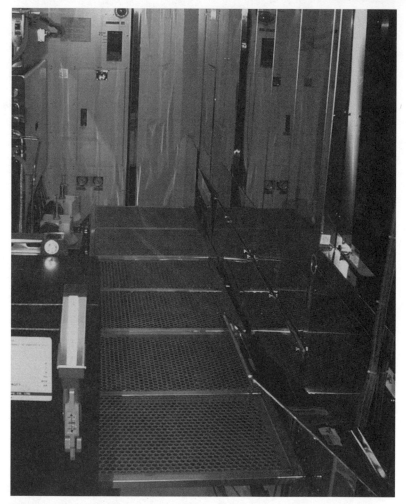

Figure 4-22. A catwalk added between two parts of a large machine

Cleaning and Inspection Materials

Agilent found that the best way to support team cleaning and inspection activities was to provide a cart containing all of the tools and supplies that the team needed to perform its work. This cart contains simple tools, cleaning chemicals, cleaning wipes, mirrors, flashlights, gauge-marking materials, and other supplies to aid the team. Figure 4-23 shows a typical cleaning and inspection cart.

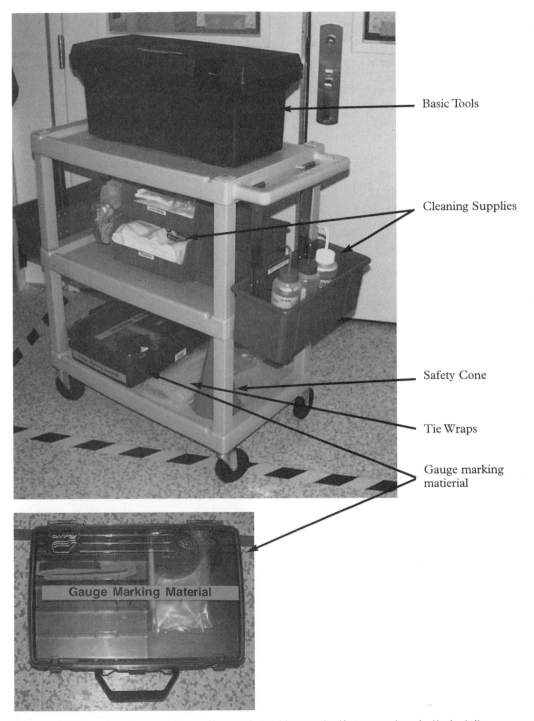

Basic Tools

Cleaning Supplies

Safety Cone

Tie Wraps

Gauge marking matierial

Gauge Marking Material

Figure 4-23. A cleaning and inspection cart used by production operators in their daily cleaning and inspection maintenance work

M-Tags

M-Tags, or Maintenance Tags, are used as a system of tagging minor defects found during equipment inspections so that they can be repaired. Companies use many different types of M-Tag systems (also known as F-Tags, the F standing for fuguai, the Japanese word for abnormality). In many work environments, a tag can be attached and left directly at the defect on the equipment.

Figure 4-24 shows such a tag used on equipment at Agilent.

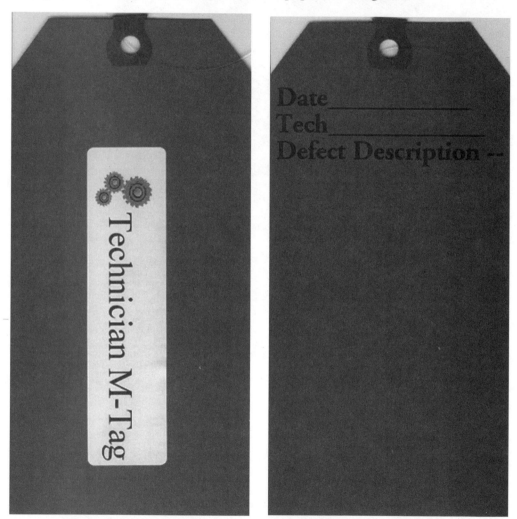

Figure 4-24. A simple M-Tag used on equipment outside of Agilent's clean room

However, such tags cannot be used on equipment inside the IC clean room, so Agilent developed a different M-Tag system for this equipment. Figure 4-25 shows such a tag. Teams use instant cameras to take pictures of the defects and attach the photos to an M-Tag with an explanation of their finding. The M-Tags

HEWLETT® PACKARD **M-Tag**

Check One:
- Safety
- Quality
- 5S
- X Repair
- Improvement
- Other

Machine *Oxide Dep #10*

Team Name *B-Diff* Date *12/12/97*

Originator *Jeff Brooks*

Problem photo or sketch–

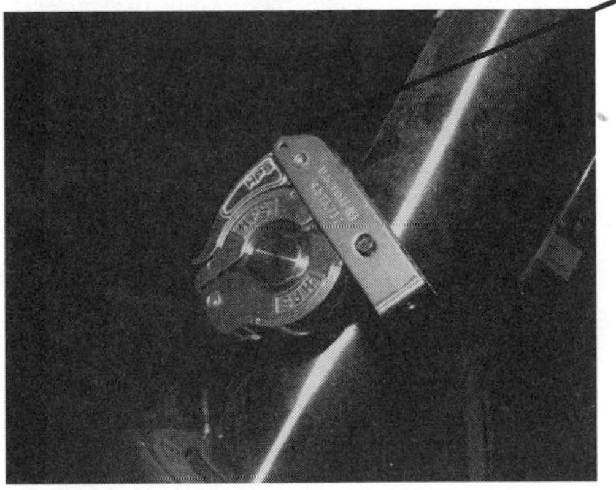

Problem: Missing cotter pin on main vacuum line

Countermeasure status: *Missing Cotter Pin Replaced*

Work to be done by: ___ Production *X* Maintenance ___ Engineering

Person in Charge *Mike Mora* Target Due Date *12/24/97* Completed Date *12/14/97*

M-Tag Closed by *Jeff Brooks* Date *12/18/97*

Figure 4-25. A typical M-Tag produced for defects found inside Agilent's clean room are then posted on a technician's M-Tag assignment board, outside the clean room, as shown in Figure 4-26. Any technician working in that machine's area on any shift can resolve any M-Tag hanging on the board. Agilent's expectation is that no M-Tag shall be left open for more than two weeks without being repaired unless a sound reason for its noncompletion is offered, such as awaiting a parts order.

Open M-Tags are transmitted from operators to techs by hanging them on the appropriate tech board

Closed M-Tags are transmitted from techs to operators by placing them in the appropriate team box

Figure 4-26. M-Tags produced inside the clean room by operators are deposited on the appropriate tech group clipboard on this M-Tag organizer. Techs on any shift from that group repair the machine defects and return the completed M-Tags to the originators.

1-Point Lessons

Many equipment improvements are the result of improvements in people's activities in operating and maintaining their equipment. One way team members can share their learning experiences with one another and with other teams is by using 1-Point Lessons.

People need a learning system that complements the nature of machines: complex machines are assemblies of very large numbers of fairly simple components; a 1-Point Lesson system has the same nature—thousands of lessons, each about one simple thing.

To create high-quality, usable 1-Point Lessons, Agilent generated the following guidelines.

- Each lesson contains only a single theme to be learned. (1-Point Lessons are not suitable for long lists of detailed instructions. Those are maintenance or operating procedures.)
- Each lesson should fit onto one page.
- A lesson may describe a basic skill, an equipment problem that people need to be aware of, or an equipment improvement that has been made.
- A lesson should contain more visual information than text.
- Text should be straightforward, easy to understand, and go straight to the point.
- A lesson can begin by explaining the need for this knowledge, which may include what, why, when, how, or cautions.
- A lesson should be able to be easily learned by its intended audience in less than 5 to 10 minutes. If it is not, it must be broken down into its constituents.
- Once lessons are generated, they are distributed to their intended audience.
- Those who have already learned them teach lessons to new people.
- Lessons are best taught *at the machine*, with the 1-Point Lesson in hand.
- Once the lesson has been learned by its entire intended audience, it is retained in the appropriate document location:
  - Machine-specific 1-Point Lessons become part of that machine's operating or maintenance procedure.
  - Generic "technology" lessons—about fasteners or bearings, for instance—go into Agilent's *Basic Machine Technology Manual*, which contains chapters for every general technical subject.

At Agilent, we expect operators, techs, and engineers to create 1-Point Lessons every time they learn something new that is useful to others. Figures 4-27 through 4-29 are examples of 1-Point Lessons produced by Agilent's operators and technicians.

1-Point Lesson

Equipment *Coat*

Team *C-Photo*

Date *11/26/98*

Created by *Bill Reinhart*

Eng/Tech Approval *jal*

Theme: How to check and tighten pneumatic speed control lock nuts

Pneumatic speed controls have lock nuts on them to keep machine vibration from changing the valve setting. These lock nuts need to be kept locked. To check or tighten the lock nuts:
1. Hold the adjustment knob
2. Turn the knurled lock nut clockwise until finger tight

Training Record

Date								
Trainer								
Trainer								

Figure 4-27. Example of a 1-Point Lesson Shared by operators

1-Point Lesson

Equipment   *All*

Team   *D-Dep*

Date   *7/18/99*

Created by   *Terry Brinkerhuff*

Eng/Tech Approval   *jal*

Theme: How to tighten Swagelok tube fittings

Swagelok tube fittings are used in many of our production machines. Their largest failure mechanism—leaking—results largely from misuse. There are three things every tech should know about proper tightening of Swagelok fittings:
1. First-time use—tighten 1-1/4 turns from hand tight (about 7 flats)
   (3/4 turns for tube sizes less then 1/4")
2. Re-use—tighten hand tight, then an additional 1 flat
3. Replace the ferrules when there is no clearance between the front and rear ferrules. Without clearance, the fitting will not seal.

If treated properly, ferrules should be able to be loosened and tightened indefinitely. Over-tightening is what causes them to fail.

No Clearance
=> No Good!

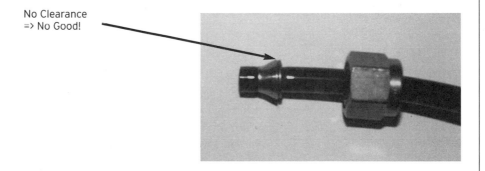

Training Record

Date									
Trainer									
Trainer									

Figure 4-28. Example of a 1-Point Lesson shared by equipment technicians

1-Point Lesson

Equipment *All*

Team *A-Dep*

Date *8/17/99*

Created by *Dave Putnum*

Eng/Tech Approval *jal*

Theme: Installing directional speed controls on pneumatic cylinders

Pneumatic cylinder speed controls are needle valves in parallel with a check valve. Set them to free flow air into the cylinder and restrict the air coming out the the cylinder. Remember this rule:

Always control the exhaust!

Installing speed controls backward will cause the cylinder to move with a jerking motion.

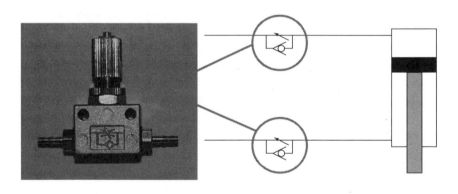

Training Record

Date									
Trainer									
Trainer									

Figure 4-29. Example of a 1-Point Lesson shared by equipment technicians

Visual Controls

Many inspections aimed at preventing minor defects in a machine can be carried out more quickly and easily if visual controls are used. The most obvious types of visual controls are those that indicate the correct position of machine gauges. Examples of those used at Agilent are shown in Figures 4-30 through 4-33. Team toolboxes contain material to make these visual controls. Technicians and engineers prescribe the correct range of the gauge positions to be identified as standard for the machine.

Green label on gauge—needle should be "in the Green Zone"

Figure 4-30. A typical visual control for most machine gauges

Red label on gauge   Green label on gauge

Figure 4-31. "In the Green! Out of the Red!" These gauges show water pressure on each side of a filter. As the filter gets dirty, the needle on the left gauge will drop into the red zone.

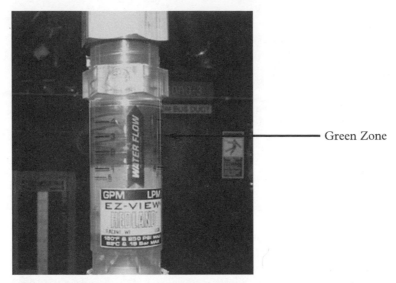

Green Zone

Figure 4-32. Typical visual control on a flow meter

Figure 4-33. A critical bolt that must remain properly torqued can be visually inspected to see if it is getting loose

Visually Controlled Inspection Routes

Some cleaning and inspection standards do not need a scheduled checklist of instructions to be followed in order for them to be completed. Many such inspections are 100 percent visually controlled and need no additional instructions. The inspection is scheduled like any other PM, but has no instruction set other than to follow the visually controlled routing system.

In the following example, equipment located outside the clean room contains vacuum pumps, RF generators, cooling-water controls, power supplies, and other equipment. There are over 40 inspection points on this equipment, but no checklist or instruction is required to complete the work. The visual control system works as follows:

1. Go to inspection station number 1 (see Figure 4-34).
2. Follow the numbered inspection points. Number 1 will always be found on your left as you face the station ID sign. Numbers 2, 3, and so on will be found as you move around the machine to the right (see Figure 4-35).
3. When you reach a number followed by an "!", you have reached the last inspection point for this station (see Figure 4-36).
4. Proceed to inspection station number 2 and repeat the above steps.
5. Continue until all of the inspection stations have been completed.

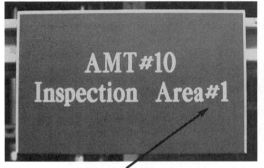
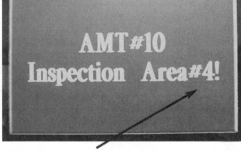

The inspection always begins at the first station.

The last station is clearly indicated by an "!"

Figure 4-34. Visual route map locators along inspection routes

Start here

Figure 4-35. The inspection always starts at the left of the station location sign and moves around the area clockwise

Figure 4-36. The last inspection point for every station is clearly indicated by an "!"

Visual Route Maps for Frequent PMs

Another kind of visual route map is used for PMs that are scheduled very frequently—daily, or perhaps weekly. In these cases, there is no need to use a PM scheduling system to generate the inspection PM. The visual controls for these frequent events can provide the scheduling as well. These visual machine controls are most often used by our production operators to schedule their maintenance work.

Agilent uses the following system of visual controls to create inspection routes—often referred to as "route maps"—for frequent visual cleaning and inspection work:

- Shape indicates frequency
- Color indicates team (shift) ownership
- Numbers indicate route map order

Agilent operates its IC fab with four shifts of operators and maintenance technicians. The shifts are either day or night, front half or back half of the week. Each shift has its own color to identify which is responsible for completing this work. The team (shift) colors are:

- A shift—yellow
- B shift—green
- C shift—blue
- D shift—red

Numbers included in the colored shapes indicate the order of completion for the route map. The last number on the route always ends with an "!".

For example, a set of **blue** circles, located at various points on a machine, indicate the PM route to be carried out by our **C-shift** team on this machine on a **weekly** basis. There are five points to be inspected or serviced on this route.

Visually Controlled Production Flow

All wafer lots are "pulled" (rather than pushed) through Agilent's IC fab with a pull-card system. The number of pull cards at each station limits the amount of WIP that can accumulate in any given area of the IC fab. Even though it could easily have been automated to let computers control the flow of production lots, the pull-card system was designed to be manual so it could provide operators with a visual image of the status and priority of every lot in the IC fab. Figure 4-37 shows a typical pull-card station. Figure 4-38 shows a typical production lot— a box of wafers with its bar-code card and pull card attached.

Figure 4-37. A manual pull-card station organizes the flow of material through the IC fab

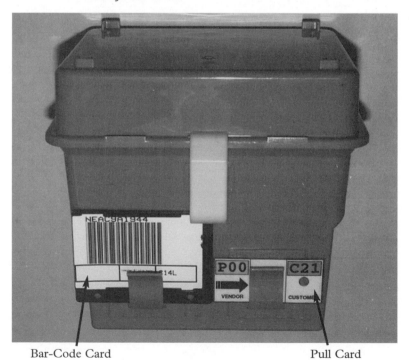

Bar-Code Card   Pull Card

Figure 4-38. A typical production lot box holding a bar-code card and a pull card

Team Notebooks and Activity Boards

Every TPM team should keep a set of notebooks to track important information gathered during team activities. This notebook should be made available to the team at the machine when they are working and should be designed to record the following information:

- The Three Lists
- Team members and the equipment they work on
- M-Tags
- 1-Point Lessons
- Team performance charts
- Before-and-after equipment photos
- A master checklist for the current TPM step activities
- Various lists for future reference:
  – Safety concerns
  – Unanswered questions
  – Difficult access areas
  – Repeated contamination areas
- Machine improvement ideas
- Team activity improvement ideas
- A place for any other information to be entered that the team desires

TPM teams should also have a team activity board—ideally located near the equipment where people work every day, but this is often impractical. At Agilent, especially, it is impossible to keep activity boards in the clean room. That is why the team works with its clean-room compatible notebooks, and then summarizes its activities on an activity board which hangs in the hallway outside of the clean room, where everyone can see it.

During Step 1, the activity board should include the following:

- The team members and their team charter
- Completed and current team activities
- Trend charts of team and machine metrics
- Any other information the team wishes to display about its activities[1]

Cleaning and Inspection Standards

Once a machine has been thoroughly cleaned and all of the humanly detectable minor defects removed, the next order of business for the team is to keep the machine in this "restored" condition. Some of the work can be performed by technicians during their PMs, but much of the cleaning and inspection work is

---

[1]For more information about activity boards, see "Team Activity Boards" on page 295.

actually better performed by production operators. Some of the inspections must be done while the machine is running, and operators are usually present during these periods of time. Other inspection work needs to be completed during the few minutes that might exist between production runs. These brief periods of time are most often unpredictable, and scheduling a technician to perform a few minutes of work at such a time is quite difficult.

Operators are much better positioned to perform these tasks than techs, but the operators need to gain the knowledge, tools, and skills required to perform this work. They also need a checklist of the work to be performed. This checklist for production operators who clean and inspect their own equipment is most often called a cleaning and inspection (C&I) standard. The standard:

- Describes the tasks to be performed
- Describes the methods and tools required
- Refers to relevant 1-Point Lessons that provide more detail about the work
- Designates the frequency at which each checklist item must be accomplished

Figure 4-39 shows a typical C&I standard for one of Agilent's machines.

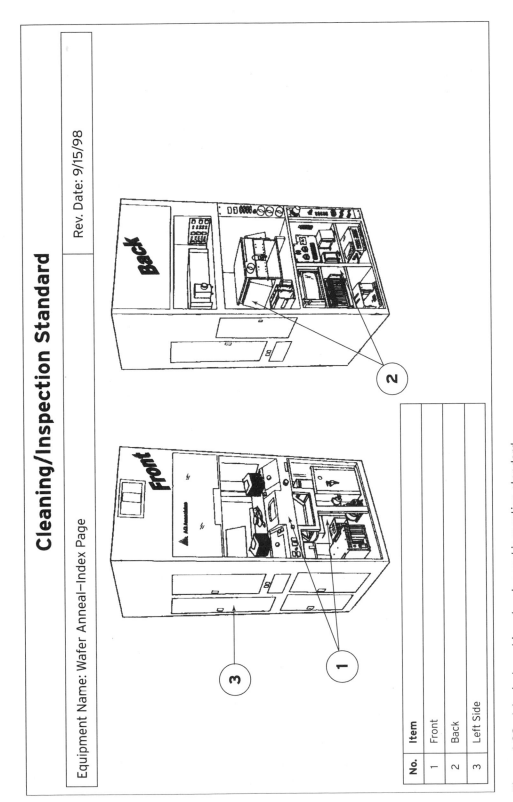

Figure 4-39. A typical machine cleaning and inspection standard

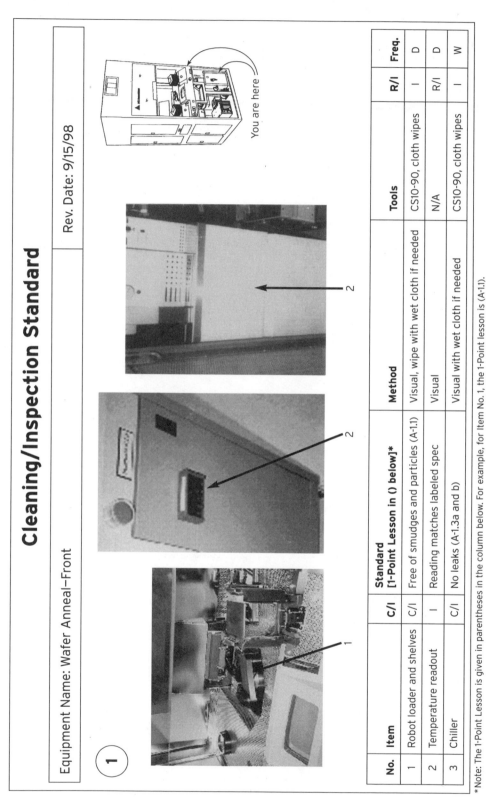

Cleaning/Inspection Standard

Equipment Name: Wafer Anneal–Front

Rev. Date: 9/15/98

You are here

No.	Item	C/I	Standard [1-Point Lesson in () below]*	Method	Tools	R/I	Freq.
1	Robot loader and shelves	C/I	Free of smudges and particles (A-1.1)	Visual, wipe with wet cloth if needed	CS10-90, cloth wipes	I	D
2	Temperature readout	I	Reading matches labeled spec	Visual	N/A	R/I	D
3	Chiller	C/I	No leaks (A-1.3a and b)	Visual with wet cloth if needed	CS10-90, cloth wipes	I	W

*Note: The 1-Point Lesson is given in parentheses in the column below. For example, for Item No. 1, the 1-Point lesson is (A-1.1).

Figure 4-39. A typical machine cleaning and inspection standard

Cleaning/Inspection Standard

Equipment Name: Wafer Anneal–Back

Rev. Date: 9/15/98

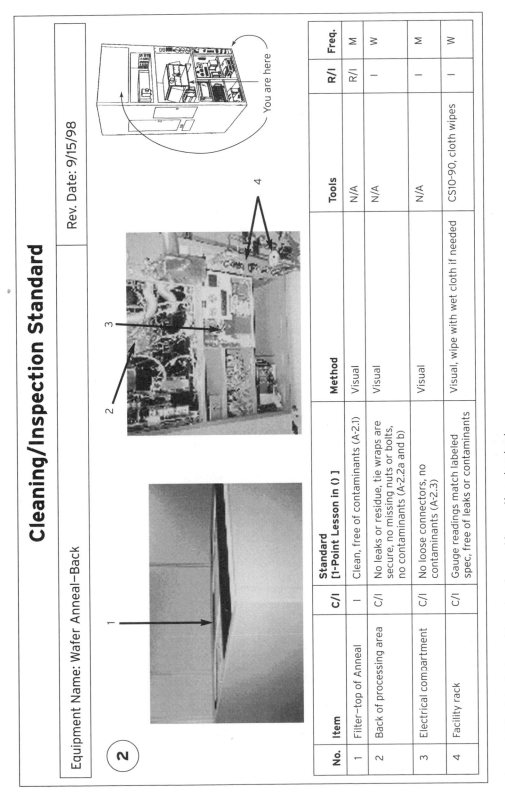

You are here

No.	Item	C/I	Standard [1-Point Lesson in ( )]	Method	Tools	R/I	Freq.
1	Filter–top of Anneal	I	Clean, free of contaminants (A-2.1)	Visual	N/A	R/I	M
2	Back of processing area	C/I	No leaks or residue, tie wraps are secure, no missing nuts or bolts, no contaminants (A-2.2a and b)	Visual	N/A	I	W
3	Electrical compartment	C/I	No loose connectors, no contaminants (A-2.3)	Visual	N/A	I	M
4	Facility rack	C/I	Gauge readings match labeled spec, free of leaks or contaminants	Visual, wipe with wet cloth if needed	CS10-90, cloth wipes	I	W

Figure 4-39. A typical machine cleaning and inspection standard

Cleaning/Inspection Standard

Equipment Name: Wafer Anneal–Left Side | Rev. Date: 9/15/98

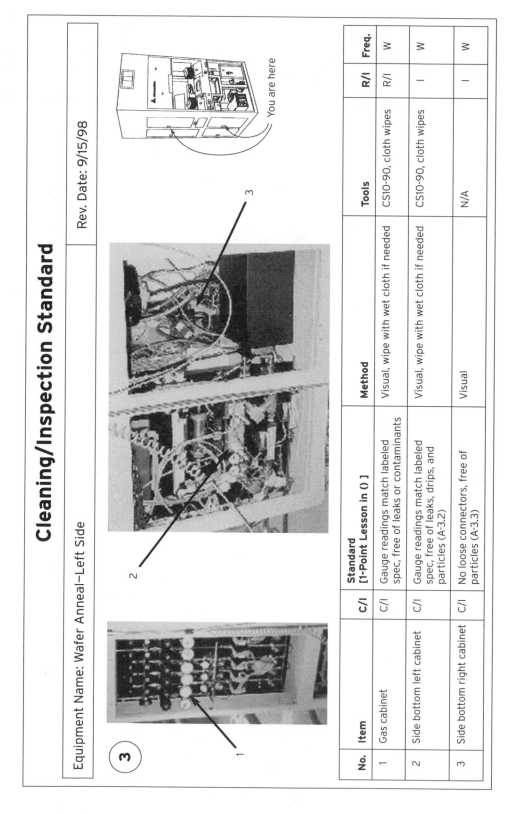

No.	Item	C/I	Standard [1-Point Lesson in ( )]	Method	Tools	R/I	Freq.
1	Gas cabinet	C/I	Gauge readings match labeled spec, free of leaks or contaminants	Visual, wipe with wet cloth if needed	CS10-90, cloth wipes	R/I	W
2	Side bottom left cabinet	C/I	Gauge readings match labeled spec, free of leaks, drips, and particles (A-3.2)	Visual, wipe with wet cloth if needed	CS10-90, cloth wipes	I	W
3	Side bottom right cabinet	C/I	No loose connectors, free of particles (A-3.3)	Visual	N/A	I	W

Figure 4-39. A typical machine cleaning and inspection standard

GETTING TPM TEAMS STARTED

A TPM team's first meeting can be a confusing experience because most team members will not have a good idea of what to do. Operators—and engineers, especially—may know little about the components of the equipment compared to what maintenance technicians know. The following sets of activities help to get a team underway on its first TPM step.

1. Begin cleaning and inspecting equipment components in a team-accessible area. Anyone can do this, even if they have no idea what the machine components do.

2. Machine cleaning is not intended to be "mindless cleaning." As people clean, they should "turn on their brains and their curiosity." They should ask other team members questions about what they are doing and seeing, such as: "How do I clean this? What should I look for? Is this part supposed to be loose? What does this do?"

    Many of the team's more technical people can usually answer these questions. If any question cannot be answered, it should be written down in the team's notebook so the answer can be discovered and shared with team members at a later date. This question-and-answer process leads to a discussion about machine components that the team members are cleaning. This discussion is the first team learning activity and the beginning of machine improvements that will result from this learning.

3. The team dialogue will lead to one of the following initial team activities:
    - M-Tags for discovered machine defects
    - 1-Point Lessons to share what one has learned about the machine with other team members
    - Entries in the team's notebook:
      - Safety concerns
      - Before-and-after photos
      - Unanswered questions
      - Sources of contamination (areas that are found dirty week after week, despite repeated cleaning)
      - Areas that are difficult to access for cleaning and inspecting
      - Ideas for improving team activities
      - The team's activity data

As soon as the team gets started, members should begin to measure their own behavior. Changes in behavior must precede improved machine performance. Behavior comes first; results come later.

The only result that the team is striving for during the early stages of TPM Step 1 is a machine that is clean and free of all humanly detectable minor defects

in accessible areas of the machine. Measurements that the team can take to indicate the progress of their activities might include the following:
- Number of M-Tags written
- Number of M-Tags closed
- Average time taken to close an M-Tag
- Number of unnecessary items removed from the machine
- Number of 1-Point Lessons written and shared
- Team meetings—hours per week
- Team meetings—percent attendance
- Team meetings—list of current activities

Once these behavior metrics indicate that progress has been made, the team can further advance its early Step 1 activities by using the checklist on the following page. This checklist describes desired team behaviors and outcomes during the first half of Step 1 activities.

> *In some respects, early TPM team activities are a "leap of faith" on the team members' part. They are engaging in time-consuming activities that don't produce an immediate and observable machine performance improvement. The link between these behaviors and improved machine performance should have been established by the TPM pilot and Manager's Model Teams so that this "leap of faith" is not too large a one for people to accept.*

ADVANCING STEP 1 TEAM ACTIVITIES

Once TPM team members have a good grasp of early Step 1 activities, they will be successfully discovering minor machine abnormalities, correcting them, and elevating their own knowledge and skill regarding cleaning and inspection.

They can now begin to move from restoration activities to improvement activities. Up to this point, little consideration has been given to the time taken to perform the team's cleaning and inspection work. To advance, the team will find ways to make their work easier and faster.

TPM ACTIVITY CHECKLIST: EARLY STEP 1 ACTIVITIES

Team _____

Equipment _____

Check each of the following items as satisfactorily completed or not.

No.	Audit Check	Complete
1	Are all team members participating in activities?	☐
2	Do all team members understand the goals of the current team activities?	☐
3	Are team activity data recorded?	☐
4	Are M-Tags created, closed, and retained?	☐
5	Are 1-Point Lessons created for relevant points?	☐
6	Are 1-Point Lessons signed off by all team members?	☐
7	Are the notebook lists being adequately used?	☐
8	Are all safety concerns immediately addressed?	☐
9	Are all questions answered?	☐
10	Have all team members developed a good eye for minor defects?	☐
11	Is the machine clean?	☐
12	Have all humanly detectable minor machine defects been eliminated on an idle machine?	☐
13	Are all unnecessary materials removed from the machine?	☐
14	Are the machine gauges visually controlled?	☐
15	Are labels and nameplates clear and legible?	☐

Cleaning Improvements

Consideration at this point should be given to sources of contamination that are found during repeated cleaning activities. Agilent teams that repeatedly had to clean the same contamination quickly tired of this job and found three ways to reduce their workload:

- Eliminate the contamination at its source so future cleanings are unnecessary
- Contain the contamination at its source to restrict the clean-up work to a smaller area
- Develop improved cleaning tools and methods to make the work easier and faster to accomplish

Maintenance Access Improvements

Many inspections can be hampered by difficult access. In one of Agilent's machines, eight bolts held a door closed. It took team members longer to unbolt the door than it took to inspect the components behind the door. The door was redesigned with a hinge and door handle, which eliminated virtually all access time for cleaning and inspection work done behind the door.

One way to think about the importance of improving cleaning and inspection access is: "If you can't access it easily, you probably won't access it very often. If you don't access it often enough, you can't properly maintain it. If you don't properly maintain it, you can't keep it from breaking down."

As teams begin to improve on their cleaning and inspection activities, cleaning and inspection work becomes more efficient. The time required to clean and inspect equipment should decrease as sources of contamination and access difficulties are improved and as team members fine tune the frequency required for each checklist item on the C&I standard (Figure 4-40).

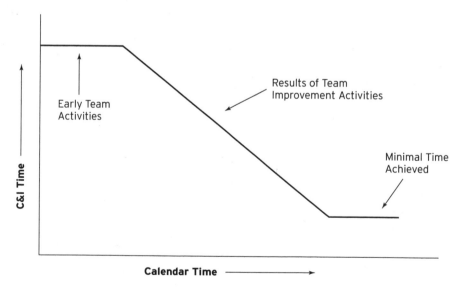

Figure 4-40. As teams make improvements, the time required to perform cleaning and inspection activities decreases

Figures 4-41 through 4-43 are examples of some machine-access improvements made in Agilent's IC fab.

Figure 4-41. This machine had large doors that provided a cosmetic cover for the machine. However, the doors had no other function and severely limited machine access in cramped quarters. The doors were permanently removed to provide improved maintenance access.

Figure 4-42. The back of an Agilent machine with its new set of clear access panels

The back of the machine in Figure 4-42 had two large stainless steel access panels. The panels were large, heavy, and awkward. Two people were generally required to wrestle either of the panels off the machine for maintenance access. This created a serious access problem, as well as a safety issue.

The two large panels were replaced with five smaller panels made of clear plastic. Visual inspections can now be made without removing any panels at all. If a panel does need to be removed, it can easily be done without tools. Maintenance access has been significantly simplified by this simple and inexpensive design change.

Seven of nine nearly identical machines in Agilent's IC fab had dark panels to protect people from the bright lights inside the machine, as shown in Figure 4-43. But most of the panels were still light enough for operators to see through so they could inspect the inside of the machine while it was running. Unfortunately, two of these nine machines had covers so opaque that operators could not observe anything that was happening inside of the machine. The opaque covers were replaced with lighter-colored covers so the machines could also be inspected while running.

Figure 4-43. One of nine similar machines with dark safety panels that protect people from bright light inside the machine

Daily Activity Scheduling

Operator cleaning and inspection work should not remain a weekly extracurricular activity; rather, a small amount of cleaning and inspection work should be done every day. Scheduling the work prescribed by the C&I standards can be done in several different ways. A production team or individual operator can be responsible for seeing that the work on a particular machine is completed on time. C&I schedules can be automated in the same way most companies automate their PM scheduling for equipment techs. Figures 4-44 through 4-46 show the checklists created from the C&I standard shown in Figure 4-39. Checklists for all of the machines taken care of by a single team are distributed so that a small amount of the cleaning and inspection work can be accomplished every day.

DEP *** DEP *** DEP *** DEP *** DEP *** DEP *** DEP *** DEP ***

OPERATOR MAINTENANCE ORDER FOR DEP

Job Name: WAFER ANNEAL DAILY

Equipment Name: Wafer Anneal

Location: DEP
Shift: Night
Flowline: C33

Date Issued: January 24, 2000

- -

This is a **DAILY** operator maintenance checklist.

The page and line numbers reference the cleaning/inspection standard for the Wafer Anneal machine. "R" items must be performed while the machine is running; "I" items must be performed while the machine is idle; "R/I" items can be performed while the machine is either running or idle.

Line Item Description	Page	Line	R/I	Initials
Robot Loader & Shelves	1	1	I	
Temperature Readout	1	2	R/I	

- Did you find anything during your inspection that requires an M-Tag?
- Do the 1-Point Lessons with the C&I standard need to be improved?
- Are there other maintenance items on the machine that should be added to the operator C&I standard?

Figure 4-44. A daily cleaning and inspection work order produced by the scheduling computer from the machine's cleaning and inspection standard

DEP *** DEP *** DEP *** DEP *** DEP *** DEP *** DEP *** DEP ***

OPERATOR MAINTENANCE ORDER FOR DEP

Job Name: WAFER ANNEAL WEEKLY

Equipment Name: Wafer Anneal

Location: DEP
Shift: Night
Flowline: C33

Date Issued: January 18, 2000

- -

This is a **WEEKLY** operator maintenance checklist.

The page and line numbers reference the cleaning/inspection standard for the Wafer Anneal machine. "R" items must be performed while the machine is running; "I" items must be performed while the machine is idle; "R/I" items can be performed while the machine is either running or idle.

Line Item Description	Page	Line	R/I	Initials
Chiller	1	3	I	
Back of Processing Area	2	2	I	
Facility Rack	2	4	I	
Gas Cabinet	3	1	I	
Side Bottom Left Cabinet	3	2	I	
Side Bottom Right Cabinet	3	3	I	

- Did you find anything during your inspection that requires an M-Tag?
- Do the 1-Point Lessons with the C&I standard need to be improved?
- Are there other maintenance items on the machine that should be added to the operator C&I standard?

Figure 4-45. A weekly cleaning and inspection work order produced by the scheduling computer from the machine's cleaning and inspection standard

DEP *** DEP *** DEP *** DEP *** DEP *** DEP *** DEP *** DEP ***

OPERATOR MAINTENANCE ORDER FOR DEP

Job Name: WAFER ANNEAL MONTHLY

Equipment Name: Wafer Anneal

Location: DEP
Shift: Night
Flowline: C33

Date Issued: January 18, 2000

This is a **MONTHLY** operator maintenance checklist.

The page and line numbers reference the cleaning/inspection standard for the Wafer Anneal machine. "R" items must be performed while the machine is running; "I" items must be performed while the machine is idle; "R/I" items can be performed while the machine is either running or idle.

Line Item Description	Page	Line	R/I	Initials
Filter (Top of Anneal)	2	1	R/I	
Electrical Compartment	2	3	I	

- Did you find anything during your inspection that requires an M-Tag?
- Do the 1-Point Lessons with the C&I standard need to be improved?
- Are there other maintenance items on the machine that should be added to the operator C&I standard?

Figure 4-46. A monthly cleaning and inspection work order produced by the scheduling computer from the machine's cleaning and inspection standard

Ongoing Cleaning and Inspection Audits

At this point, machines are clean and defect free and are being regularly maintained by a cooperative alliance of production operators and maintenance technicians. The new work created in this step should be strongly integrated into people's normal work routine. However, permanent behavior changes are hard to implement in an organization; without the proper measurement and consequences, workers will tend to slip back into old habits. It is all too possible that a year after Step 1 activities are completed, the machines will no longer be clean and defect free.

To overcome this hurdle, Agilent created a permanent machine audit process. Production and maintenance supervisors are responsible for monthly audits of the machines their people clean and inspect. In addition, a special audit team—made up of highly trained production and maintenance supervisors—audits a number of machines every month, checking them for compliance with the operator's C&I standard and technician cleaning and inspection PM's. A score is given, and combined audit scores are posted on a public trend chart. Our goal is 100 percent compliance to the C&I standard. Every team has at least one of their machines audited every month, and every machine is audited at least twice a year. This ongoing audit process has permanently maintained the clean and defect-free state of our equipment.

STEP 1 MASTER CHECKLIST

When a TPM team has completed the advanced Step 1 activities, it is ready for a Step 1 final audit. Successfully completing all of the line items on the master checklist on the next page completes the team's Step 1 activities.

STEP 1 MASTER CHECKLIST: RESTORE EQUIPMENT TO "NEW" CONDITION

Basic TPM tools have been created, are documented, and are working well:

☐ An M-Tag system

☐ A 1-Point Lesson system

☐ The Three Lists and all necessary safety modifications on all machines

☐ A system for specifying machine standards—how a component "ought to be"

☐ A TPM team structure including everyone—as much as possible—as a TPM team core member

☐ A permanent cleaning and inspection audit system

☐ The C&I standard is complete and officially documented:
  ___ The C&I standard includes inspections of machine conditions when it is operating, as well as when it is idle
  ___ Quality 1-Point Lessons are written, approved, and attached to the C&I standard for every line item
  ___ Every C&I line item is scheduled daily, weekly, or monthly
  ___ Every scheduled item is labeled: R (Inspect While Running), I (Inspect While Idle), or R/I (Inspect While Running or Idle)

☐ Every C&I line item is being completed as scheduled

☐ The C&I standard is being continually improved, and training is being completed on the improvements

☐ Every operator is competent to perform the machine cleaning and inspection activities identified on the C&I standard

☐ Cleaning and inspection items are included in scheduled tech PMs for operator-inaccessible machine areas and are included on the PM checklist

☐ All machine areas—whether accessible by production operators or maintenance techs, whether located on the factory floor or in utility areas—are clean and free of all humanly detectable minor machine defects, as shown by a history of successful audits

☐ Good teamwork is taking place between the TPM team core members and the associate members

☐ Team improvement ideas are identified and pursued

continued

STEP 1 MASTER CHECKLIST: RESTORE EQUIPMENT TO "NEW" CONDITION

☐ Teams share ideas and cooperate with one another on TPM and other manufacturing activities

☐ The time required for cleaning and inspection is decreasing

☐ Visual controls are in place on all gauges:
 ___ Proper setpoints and ranges are defined by equipment engineers
 ___ Every gauge inspection is scheduled on either a C&I checklist or a technician PM
 ___ All visual control materials are applied per visual control standards
 ___ Gauge setpoints and allowable ranges are included in the maintenance document

☐ Visual "route maps" are in place to aid in cleaning & inspection work

☐ Sources of contamination are being eliminated or improved

☐ Difficult access areas are being improved

A team activity board is in place, is being continually updated, and contains the following material:

☐ Team charter/machine ID/team members

☐ Team activities:
 ___ Identify TPM implementation progress on a machine subassembly/ step matrix[2]
 ___ Identify all activities that are completed or in progress

☐ Team results:
 ___ Metric trends
 ___ Anecdotal results

---

[2]For more information, see "Focusing TPM Steps" on page 261.

STEP 1 INFRASTRUCTURE SUPPORT

In order to carry out any step of TPM, an organization must create the supporting infrastructure to make certain team activities possible. For example, teams can't be asked to create and distribute 1-Point Lessons if there is no 1-Point Lesson system or the equipment to support it—for instance, forms, cameras, distribution lists, and the like.

An organization's existing infrastructure should be used to support TPM activities if at all possible. New infrastructure should be created only if nothing is in place that will do the job. The idea isn't to waste resources creating a new TPM documentation system if a documentation system currently exists: Use what's there. Update its capabilities as needed.

To implement TPM effectively, organizational systems need to support the following Step 1 activities and processes:

1. A process for forming cross-departmental action teams.
2. A cleaning and inspection standard process.
3. A 1-Point Lesson creation, distribution, and recording system.
4. An M-Tag system.
5. A system for scheduling line items on C&I standards at different frequencies through issued schedules or by visually controlled "route maps."
6. A location to record gauge setpoints and allowable ranges.
7. An activity board process.
8. Systems for producing and displaying performance metrics used by teams and managers.
9. Ongoing manager audits of machine compliance to the cleaning and inspection standard and similar technician PMs.
10. Changes in evaluation, ranking, and reward systems to support wanted behaviors and to stop rewarding old, unwanted behaviors. For example, reward those people who take an active role in the continual elimination of minor machine defects.

A similar list of organizational infrastructure requirements is included with every TPM step checklist in this book. When creating these systems, consider all of the following for each:

1. Who is affected by this system?
2. What forms or process flowcharts are required?
3. Where should the information be placed?
4. How do people access it?
5. What is the approval process for using this procedure?

6. What is the process to update the information (as we learn how to improve it)?
7. What training system will be used to teach people how to use this information?

STEP 1 DELIVERABLES

When TPM Step 1 has been completed on a machine, the machine should be running considerably better than it did before this step began. The clean and defect-free state attained on the machine should also now be continually maintained as a routine part of daily operator and maintenance tech procedures so the machine will never regress to its original level of defects. By now operators and technicians understand the importance of eliminating minor defects and keeping every machine component "as it should be."

However, keep in mind that Step 1 restoration to "like new" is only the beginning of the total machine restoration process. The fact is, many minor defects almost certainly remain in the equipment but are beyond human-detection capabilities at the access level provided for the team during Step 1 activities. Later TPM steps will uncover defects residing at a deeper level.

STEP 1'S MINDSET CHANGE
Minor equipment defects—once thought to be of no importance to machine maintenance—are now considered to be the root cause of almost all machine failures and are meticulously kept out of factory equipment. This requires operator involvement in machine cleaning and inspection, once thought to be the sole province of maintenance technicians. It also requires technician involvement in keeping minor defects out of equipment areas not accessible to operators—a task not previously performed by most equipment techs.

5

Step 2: Identify Complete Maintenance Plans

This chapter describes how to implement
TPM Step 2, including:

- Identifying the seven elements required for a complete
  scheduled maintenance plan
- Designing maintenance plans

STEP 2 GOALS

In Step 1, activities were performed to restore equipment to "new" condition by making all areas of the equipment clean and free of humanly detectable minor defects.

Step 2 describes how to identify, organize, and schedule a preventive maintenance (PM) plan. The plan for each machine is developed from the machine vendor's recommended maintenance plan and a team's own experience with maintaining the machine. Quality tests that currently need to be performed on the machine are also identified.

IDENTIFYING A COMPLETE MACHINE PREVENTIVE MAINTENANCE PLAN

A complete scheduled maintenance plan consists of seven elements:
- PM checklists
- PM schedules
- Inspection specifications
- Replacement part numbers
- PM procedures
- Part logs
- Quality checks

For the preventive maintenance plan to be effective:

- The seven elements must be in a common form and location for all equipment and must be made available to maintenance technicians through the same medium.
- The plan and each of its seven elements need to be organized by the "5S" principles: sort, stabilize, shine, standardize, and sustain the workplace organization. In other words, there is a place for each item, and each item is in its place.
- Any maintenance technician must be able to access any piece of the maintenance plan for any machine immediately, without searching for its pieces.

PM Checklists

To help organize the procedures included in the maintenance plan, each preventive maintenance (PM) procedure is given a name; that name is used as the title for a checklist of items to be completed during the scheduled PM. For example, PMs scheduled for one of Agilent's machines include the following:

- System Biweekly Inspection PM
- System Lubrication PM
- CVD Chamber Throttle Valve PM
- CVD Chamber Isolation Valve PM

Each of these PMs is scheduled by our central maintenance scheduling system, and the details of the PM are provided on a PM checklist. Figures 5-1 through 5-3 show sample checklists for some of these PMs.

Machine vendors generally supply some kind of recommended maintenance plan for their equipment. Maintenance techs may already be using all or part of these plans. They may also have altered the maintenance plan based on their experience with maintaining the machine.

Unfortunately, Agilent's experience with most existing maintenance plans is that they either over- or undermaintain equipment components as much as 80 percent of the time. Excess effort and expense are required of the maintenance department, yet machines continue to break down. Obviously such a maintenance plan needs improving. In Step 2, we will identify, organize, and schedule the PM plan. Further improvements will be made in later TPM steps.

One simple way to identify elements missing from the current PM plan is to examine the history of machine failures in order to identify the need for preventive maintenance work. For example, one of Agilent's machines had heat lamps that failed about once every month. The lamps were on continuously and had a life rating of 775 hours; they failed after living an expected natural lifetime. Unfortunately they often burned out in the middle of processing a large number of wafers, causing some of these very expensive parts to be scrapped.

System Biweekly Inspection PM

Item No.	Checklist Item	Specs	Part Numbers	Procedure Reference
1	Check all CVD chamber ballast pressures	35–45 mT	N/A	Maintenance Screen*
2	Check purge flows to the etch turbo pumps			Agilent Sputter Etch Chamber Manual, Section 3.1
	• Foreline ballast	35–45 mT	N/A	
	• Bellows bias	Add 15 mT	N/A	
3	Run AFC flow cal	25–35 mT	N/A	Agilent Gas System Manual, Section 1.4
4	Perform rate-of-rise on load lock chamber	< 20 mT/min	N/A	Maintenance Screen*
5	Set the load lock purge pressure	350–400 mT	N/A	Agilent Load Lock Chamber Manual, Section 6.6
6	Check all cooling fans: • VME rack (3) • Controller (3) • RF matches (2 each) • Mini-controller (3) • Main frame (2)	Fan blades turning and noticeable air flow	N/A	N/A
7	Check heat exchanger water resistivity	> 4 megohms	N/A	Maintenance Screen*
Evaluate this PM. How can it be made easier, faster, and better?				

*Note: "Maintenance Screen" means that the procedure can be found on the machine's CRT in maintenance mode.

Figure 5-1. A typical PM checklist

CVD Chamber Throttle Valve PM

Item No.	Checklist Item	Specs	Part Numbers	Procedure Reference
1	Remove the entire throttle valve assembly and replace with a rebuilt unit (1), including the slit window (2), UV filter (3), and O-ring (4)	N/A	1) 430-6850 2) 499-1066 3) 430-1134 4) 430-1587	Agilent CVD Chamber Manual, Sections 2.4–2.5
2	Replace the following O-rings: • Gas inlet (2) • Slit valve • Chamber lid	 N/A N/A N/A	 430-1529 430-1583 430-1549	Agilent CVD Chamber Manual, Section 2.7
3	Inspect the lamp quartz window thickness. Replace if out of spec.	50% of original thickness	499–2331	Agilent CVD Chamber Manual, Section 2.8
4	Clean and lube wafer and susceptor lift lead screws	N/A	430-2782 (Torr-Lube)	Clean with ISO. Apply light coating to entire lead screw.
Evaluate this PM. How can it be made easier, faster, and better?				

Figure 5-2. A typical PM checklist

CVD Chamber Isolation Valve PM

Item No.	Checklist Item	Specs	Part Numbers	Procedure Reference
1	Check condition of the throttle valve belts (1), the susceptor belt (2), and the wafer lift belt (3). Replace if any abnormalities are found	No surface cracks, abnormal tooth wear, or fraying	1) 490-1367 2) 490-1162 3) 490-1162	Visual
2	Replace the following O-rings: • Isolation valve poppet • Isolation valve bonnet	N/A N/A	430-1837 430-1510	Agilent CVD Load Lock Manual, Section 4.3
Evaluate this PM. How can it be made easier, faster, and better?				

Figure 5-3. A typical PM checklist

There wasn't anything wrong with the heat lamps, but we had never bothered to schedule their replacement. We waited for them to fail before changing them. A simple monthly PM was scheduled to replace these lamps, which prevented all lamp failures while utilizing 94 percent of the lamp's expected life. Since the lamps are relatively inexpensive, this preventive maintenance allows us to change the lamps when it is convenient for us, and they no longer burn out during wafer processing.

Assemble an initial maintenance plan using the machine vendor's recommended PM plan and your own experience with repairing and maintaining the machine.

PM Schedules

Once the PM checklists have been created, each PM must be scheduled with a single maintenance scheduling system. The plan will not be effective if multiple systems are used. When our Agilent pilot team began Step 2 on our TPM pilot machine, we found only a few PMs scheduled on our primary PM scheduling system. Other PMs were being scheduled by a laptop computer kept behind one of the machines, our process quality tracking system, individual computer systems, and notes technicians posted on the walls in their offices.

When we attempted to place the entire PM plan for this machine "on the table," we found that we could not easily identify all its pieces. In fact, four months after we believed we had entered all the PMs into our primary maintenance scheduling system (and thought we had "killed off" all the spurious PM scheduling tools), a PM triggered by someone's personal computer program came due in the middle of the night. The machine was down all night before someone realized there was a mistake; the PM no longer needed to be done at that time.

Agilent now uses a single, computerized maintenance scheduling software package to schedule all PMs. They are scheduled as either time-based or machine use-based. Condition-based PMs have inspections scheduled as either time- or use-based.

 If you cannot see your entire maintenance plan clearly and plainly for every machine, you need to improve the organization and visibility of your machine maintenance scheduling plans.

Figure 5-4 shows a typical Master PM Plan for one of Agilent's production machines. Each PM is scheduled by number of weeks, number of wafers processed by the machine, or the number of hours of use on a machine meter—in this case, each chamber's RF power supply. An estimate of the number of days remaining before a use-based PM is due is calculated by the rate the PM interval is being consumed. This provides technicians with an estimate of when a use-based PM will come due. The estimate improves in accuracy as the PM gets closer to being due.

It is not necessary to be a high-tech company or own maintenance-scheduling computers to effectively manage preventive maintenance. Some companies use paper-based systems such as the following, and do quite well with them:

- PMs are scheduled by a tickler file using index cards, one for each PM.
- PM checksheets and machine service data—such as part logs or condition-measurement trend charts—are kept in an organized set of folders in a file cabinet.
- Maintenance procedures are kept in an organized set of books in bookcases.
- Data for improvement metrics is collected by "tic marks" on a wall chart or clipboard at the machine.

As long as the maintenance planning system is highly organized, readily visible, available to all people using it, and contains all the important maintenance plan elements, using a computer for scheduling is not a requirement.

Some maintenance technicians believe equipment failures can be used for scheduling maintenance work, but this is *breakdown* maintenance. For example, many consumable parts in our machines used to be "scheduled" for maintenance when they deteriorated to the point where the machine would no longer operate properly. When the machine could no longer be run, operators called techs to replace the worn parts. This type of "scheduling" is a difficult habit to break; even today, Agilent techs and managers remain on guard against this maintenance practice.

Oxide Dep #10
Equipment Master PM Plan
3/29/00

PM Name	PM Interval	Interval Unit	Last PM Date	Last PM Meter Value	Current Meter Value	Next PM Due @	% Interval Completed	Days Until PM Due
System								
System Biweekly Inspection PM	002	Weeks	3/29/00	0	0	4/12/00	0%	14
System Software Backup PM	008	Weeks	2/19/00	0	0	4/15/00	70%	17
System Quarterly Safety PM	012	Weeks	3/5/00	0	0	5/28/00	29%	60
System Annual Safety PM	052	Weeks	4/7/99	0	0	4/5/00	98%	7
Remote Frame Gauge Inspection PM	002	Weeks	3/29/00	0	0	4/12/00	0%	14
Remote Frame Detailed C&I PM	012	Weeks	1/16/00	0	0	4/9/00	87%	11
Remote Frame Lubrication PM	052	Weeks	9/28/99	0	0	9/26/00	50%	181
Remote Frame Filter PM	052	Weeks	9/2/99	0	0	8/31/00	57%	155
System Wafer-handling PM	16000	Wafers	3/29/00	88258	88312	104258	0%	0
System Lubrication PM	96000	Wafers	11/5/99	53763	88312	149763	36%	258
AMMI: C&I Monthly *	28	Days	3/19/00	0	0	4/16/00	36%	18
AMWI: C&I Weekly *	7	Days	3/26/00	0	0	4/2/00	43%	4
Chamber A								
CVD Chamber Wet-Wipe PM	425	RF_HRS	3/4/00	9453	9764	9878	73%	9
CVD Chamber Isolation Valve PM	850	RF_HRS	2/4/00	9122	9764	9972	76%	17
CVD Chamber Throttle Valve PM	2800	RF_HRS	11/5/99	8133	9764	10933	58%	104
CVD Chamber Lift Assembly PM	7000	RF_HRS	3/17/99	5660	9764	12660	59%	267
CVD Susceptor PM	800	RF_HRS	3/4/00	9453	9764	10253	39%	39
CVD Chamber Slit Valve PM	4000	RF_HRS	9/15/99	7582	9764	11582	55%	163
QUAL **	3	Days	3/27/00	0	0	3/30/00	73%	1

PM Name	PM Interval	Interval Unit	Last PM Date	Last PM Meter Value	Current Meter Value	Next PM Due @	% Interval Completed	Days Until PM Due
Chamber B								
CVD Chamber Wet-Wipe PM	425	RF_HRS	3/2/00	8369	8702	8794	78%	7
CVD Chamber Isolation Valve PM	850	RF_HRS	1/22/00	7893	8702	8743	95%	3
CVD Chamber Throttle Valve PM	2800	RF_HRS	11/5/99	7047	8702	9847	59%	100
CVD Chamber Lift Assembly PM	7000	RF_HRS	3/24/99	4755	8702	11755	56%	287
CVD Susceptor PM	800	RF_HRS	1/22/00	7893	8702	8693	101%	-1
CVD Chamber RF Cal PM	2400	RF_HRS	12/8/99	7396	8702	9796	54%	94
CVD Chamber Slit Valve PM	4000	RF_HRS	9/30/99	6700	8702	10700	50%	181
QUAL **	3	Days	3/28/00	0	0	3/31/00	37%	2
Chamber C								
Etch Chamber Rebuild PM	800	RF_HRS	2/6/00	3469	3789	4269	40%	78
Etch Chamber Throttle Valve PM	1600	RF_HRS	2/6/00	3469	3789	5069	20%	208
Etch Chamber Slit Valve PM	2000	RF_HRS	11/5/99	2952	3789	4952	42%	201
QUAL**	1	Weeks	3/23/00	0	0	3/30/00	87%	1
Chamber D								
Etch Chamber Rebuild PM	800	RF_HRS	11/5/99	3119	3789	3919	84%	28
Etch Chamber Throttle Valve PM	1600	RF_HRS	3/17/99	2209	3789	3809	99%	5
Etch Chamber Slit Valve PM	2000	RF_HRS	11/5/99	3119	3789	5119	34%	288
QUAL **	1	Weeks	3/28/00	0	0	4/4/00	15%	6

* Denotes operator cleaning and inspection maintenance. (The AM in the name stands for Autonomous Maintenance, the TPM term often used to describe operator maintenance.)
** Denotes a quality monitor run by operators to check the condition of the process quality in this machine chamber.

Figure 5-4. A typical master PM plan for a production machine.

Do not use equipment failure to schedule maintenance work. This is breakdown—not preventive—maintenance.

Designing Maintenance Schedules

Graceful Deterioration Graceful deterioration means that a measurable aspect of a machine part or its performance indicates the onset of "old age" and its eventual failure. Figure 5-5 illustrates this concept.

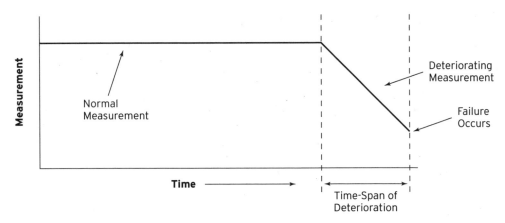

Figure 5-5. Illustration of graceful deterioration of a machine part

The measurement indicating component deterioration may be made by human senses, such as a visual observation, or it might be measured by some kind of instrument. For example, the deterioration in a power transfer belt is fairly easily observed.

Standards for the component can be set—such as the number of cracks per inch allowed, or the specification that no cord be visible through the side of the belt. As a component deteriorates, previous experience with its deterioration and failure can actually make the timing of its failure predictable.

If the pattern of decline and the time-span of the deterioration are known, a measurement history of decline can be used to predict remaining life. This allows for scheduling the part's replacement—often allowing time to acquire the part if necessary, making spare part inventories smaller.

In some cases, the measurement may be beyond the capacity of human senses. For example, the deterioration of a bearing can be measured by vibration or ultrasonic measurement, the data collected and plotted over time, and the state of deterioration thus easily visualized.

Parts that deteriorate gracefully and have a deterioration time-span of significant length are very easy to maintain with condition-based maintenance. The maintenance scheduled for these parts is not part replacement but part inspection. Measurements are taken at scheduled intervals to discover the part's state of deterioration. Once these measurements indicate that the part is deteriorating toward failure, a service or replacement can be scheduled. To prevent failure, the inspections need to be scheduled more frequently than the time-span of the detectable deterioration.

Many parts in a production environment behave this way and have significantly long deterioration time-spans. Condition-based maintenance is preferable for these parts because it allows each individual part to function for the duration of its maximum useful life.

Condition measurements may be collected in any number of ways:

- Inherent tell-tale signs that are humanly detectable, like the tread remaining on an automobile tire.
- Installed monitors that continually measure a condition, like temperature and pressure gauges installed on a machine.
- Scheduled monitoring that is performed as a PM activity, such as inspecting the level of water in a tank or measuring the vibration signature of a bearing.
- Built-in testing, such as a ground-fault interrupt (GFI) circuit breaker that provides an instant performance check at any time without external test equipment.[1]

Nongraceful Deterioration Some machine parts, such as light bulbs, do not deteriorate gracefully but fail suddenly, without warning. There are no easily measured parameters on a light bulb that will allow us to predict when the bulb will fail. However, many parts such as these have predictable life expectancies. This is especially true in a manufacturing environment where machines reside in the same environment day after day and perform the same task over and over.

Often, the largest variable in equipment life expectancies is how people maintain the parts. When they are properly maintained, these parts have reasonably predictable life expectancies. They can be serviced with either time-based maintenance or use-based maintenance.

Time-based maintenance is a maintenance plan scheduled by calendar time, such as changing the oil in a car engine every three months. Machine components in continual or consistent service can easily be scheduled on time-based maintenance because the calendar deterioration rate is fairly constant.

Use-based maintenance is maintenance scheduled by some measurement of machine use, such as hours of actual use or, in keeping with the previous example, changing the oil in a car engine every 3,000 miles. Machines or machine components used intermittently and unpredictably are best maintained with use-based maintenance, utilizing some kind of meter that correlates well to their parts-deterioration rates.

---

[1] For more information, see "Continuous Condition Monitoring" on page 241.

For parts that have a somewhat predictable life expectancy, the part replacement interval can be identified, depending on a few variables:

1. The standard deviation of the life of the part.
2. The safety interval required to prevent any part failure, depending on:
   - The cost of the failure
   - The cost of the PM to prevent the failure

All of these considerations can be weighed to determine an economically acceptable part-replacement interval. This concept is illustrated in Figure 5-6.

Note: A part life expectancy curve with a very wide deviation often indicates that parts are being subjected to accelerated deterioration, and therefore their life spans are less predictable.[2]

When a part with unpredictable life expectancy suddenly fails, perform an economic evaluation to determine how this part can be maintained:

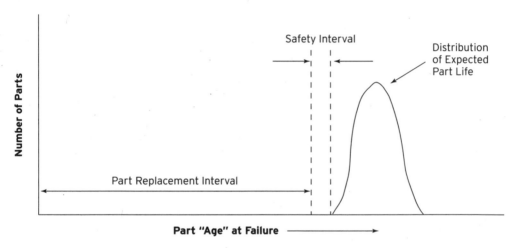

Figure 5-6. The concept of predictable life expectancies

1. If the failure has minimal economic impact and can be quickly repaired, then the part can be kept on a breakdown maintenance plan. In this case, no effort will be made to schedule a part replacement. Rather, the spare parts, tools, and procedures to replace the part will be provided, and the part will be serviced only after it has failed.

   An example of this type of part at Agilent is a computer monitor used on some machines. The monitors receive no maintenance and often die with

---

[2]For more information on narrowing the distribution of a failure curve, see "Condition-of-Use and Life Analysis" on page 222.

little warning at an unpredictable age. We keep spare monitors on hand, and these parts can be replaced within a few minutes with minimal disruption to the manufacturing line.

2. If the failure has significant economic impact, then no acceptable maintenance plan can be made to accommodate this situation. Either the part will be allowed to fail, and the economic consequences of the machine breakdown tolerated, or the subassembly in the machine that contains the part will have to be *redesigned* so that it can be maintained. Equipment redesign may include continuous condition monitoring.[3]

Machine Redesigns There are two primary situations where machine redesigns are desirable. One is the case just described, where a satisfactory maintenance plan is not possible to prevent machine failure, given the current design. The second situation is when it is important to extend the part replacement interval. This does not reduce machine failure rates, but instead reduces the amount of maintenance required to prevent the machine from failing.

For example, suppose a part with a one-year life was "beefed up" so it could last five years instead. The replacement interval for this part would be increased by a factor of five. This reduces the number of replacement parts required, the amount of technician time used each year to replace the part, and the amount of time the machine is out of production having this part changed.

One set of machines in Agilent's IC fab contained over 300 pneumatic valves, each requiring lubrication to achieve a natural lifetime. However, we could not lubricate these valves in the clean room because of contamination concerns, so they lasted only about one year each. The valves gave warning of their impending failure several weeks in advance, so we were able to replace them before they failed. However, this meant, on average, that we replaced about half a dozen of these valves every week. This was so time-consuming that a valve redesign was justified. New valves, which contained no seals and had no lubrication requirements, were designed into the machine, eliminating all valve maintenance. Following the redesign, none of the new valves failed during a five-year period. This machine redesign saved a tremendous amount of maintenance work.

Figure 5-7 summarizes the maintenance plan development process for a machine once all the deteriorating components have been identified.

 Attempting to develop a maintenance plan for each component is always our first choice. Machine redesign occurs only if an acceptable maintenance plan cannot be developed.

---

[3]For more information, see "Continuous Condition Monitoring" on page 241.

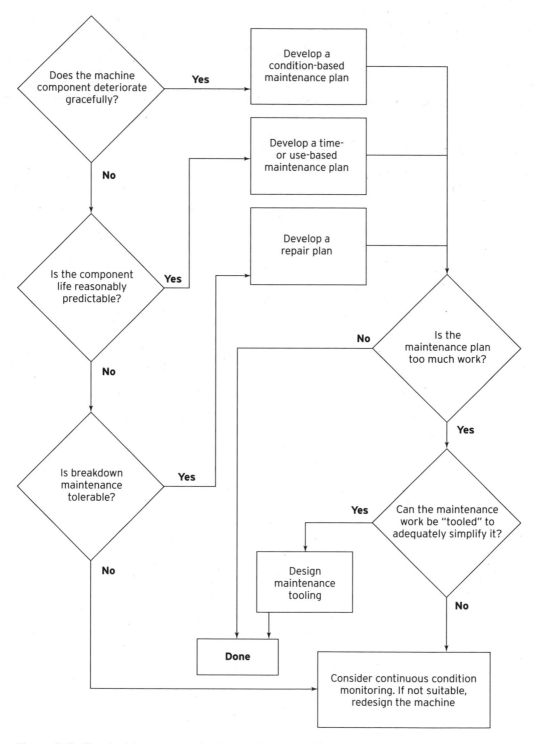

Figure 5-7. The decision process for developing a machine maintenance plan or equipment redesign

Inspection Specifications

Specifications for PM inspections must be clearly and objectively stated. An instruction such as "Inspect the quartz window and replace if needed" is inadequate because "if needed" is not defined. In this particular example, the window is good until 50 percent of its thickness is etched away. The specification column in the PM checklist includes this type of information. Figures 5-1 through 5-3 provide examples of typical specifications provided on PM checklists.

Replacement Parts

Part numbers for every part to be replaced should be in the maintenance technician's hands before the PM even begins. Therefore, the PM checklist must provide part numbers of all parts to be replaced during the PM so that they can be collected before the machine is taken down for a PM, rather than after. This is especially important in a clean room because the parts stockroom is outside the clean room. Much time and effort is wasted if techs have to run in and out of the clean room, getting into and out of their clean-room gowns, every time they discover they don't have all the parts needed to perform a scheduled PM. See Figures 5-1 through 5-3 for examples of part numbers provided on PM checklists for scheduled replacement parts.

PM Procedures

For each PM checklist item requiring more than a generic operating or maintenance procedure, a detailed written description must be provided. It does little good to schedule a complex PM if technicians are not provided with a useable procedure for completing it correctly. Otherwise, its completion will vary from time to time and person to person.

On the other hand, techs who repeat a PM frequently may not need a written procedure in front of them every time they perform the PM. But the procedure does need to exist and be referenced on occasion to prevent people "drifting" from it. The written procedure for each PM checklist item should be referenced on the PM checklist. See the PM checklist samples in Figures 5-1 through 5-3 for examples of PM procedure reference. Procedure documentation is the basis for precision maintenance, which is addressed in TPM Step 3.

Simple procedures are most often described in text form. More complex procedures are inadequately described by text and require higher levels of description, such as visual documentation.[4] Figure 5-8 is an example of a simple, text-style maintenance procedure. Any Agilent technician with basic training on this machine can easily follow the instruction provided.

---

[4]More complex procedures often require visual documentation, as described in "Precision Documentation" on page 144.

Baratron Cross Check on Chambers A & B

1. Use this procedure to determine if the Baratron pressure sensors on Chambers A & B provide identical readings of pressure.
2. Put the machine control system in the Manual Mode with Chambers A & B "OFF LINE."
3. Check that Chamber A & B susceptors are at the same temperature.
4. Run a leak-back test on both chambers. Ensure that both pass before continuing (< 20 mT/min).
5. On the Control Handler Screen, select "Ignore Vacuum Sensors."
6. Close the isolation valves on the Loadlock and Chambers A & B.
7. Open the Chamber A & B slit valves.
8. Set the O_2 gas flow on either chamber to 200 sccm but do not turn on the gas.
9. Momentarily turn on the O_2 gas to raise the pressure in the system to about 1 Torr. Monitor each chamber's Baratron reading on either the Big I/O screen or the General Charting screen. Record Chamber A and Chamber B Baratron readings at this pressure.
10. Continue releasing small amounts of O_2 into the system to achieve the following approximate pressures: 5, 10, 20, 40, and 60 Torr. Record the Chamber A and Chamber B Baratron readings at each of these pressures.

Nominal System Pressure	Chamber A Baratron Reading	Chamber B Baratron Reading
1 Torr		
5 Torr		
10 Torr		
20 Torr		
40 Torr		
60 Torr		

11. If the chamber Baratrons read within 2% absolute pressure of one another over the entire test range, they can be considered accurate.

Figure 5-8. A typical text maintenance procedure

Part Logs

Effective maintenance planning often requires knowing the age of numerous machine parts. Maintenance scheduling systems keep track of maintenance work to be done in the future, and also document past PMs. These are the only records required to learn the age of many of the parts being replaced during these PMs. For instance, suppose a PM calls for a certain part to be replaced, and every time this part is replaced, a new O-ring is added. The age of the O-ring in this part can be determined from records of when the last PM was done. This is true even if the PM was unscheduled for some reason—a breakdown, perhaps. But suppose this is a monthly PM, and the O-ring only needs to be replaced annually. A history of completed PMs will not provide the age of the current O-ring. It might be determined by searching technician PM notes, but it might not.

There are two ways to track the age of replacement parts in a situation like this. The first is to keep a separate PM for the O-ring; the only item on the PM checklist would be the replacement of this O-ring. That way, if it were replaced prematurely because of some unexpected mishap, the replacement PM would be completed prematurely and the PM log would record the time of the PM; thus, one could learn the O-ring's age. However, this approach leads to the creation of many single-item PMs for the sole purpose of keeping track of the age of critical parts.

The second—and preferred—way to track the age of replacement parts is to use a part log. Agilent has built a part log into its computerized maintenance scheduling system. Every part that needs to be tracked is entered into the part log, and the date (or meter value) noted when the part is replaced. So a monthly PM to service a part might simply state: "Replace the O-ring annually." A quick lookup in the part log of this machine will provide the O-ring's age. If the O-ring needs to be replaced, it can be pulled from stock before the PM is started. If unexpected damage is discovered during the PM and the O-ring is replaced ahead of schedule, the part log is updated, and the one-year age count begins from the replacement date.

A part log is also useful for collecting data on parts that do not have a known life expectancy. This data can then be analyzed to determine the actual life of the components. If the life distribution has a wide deviation, the part's necessary conditions-of-use can be defined, investigated, and maintained to make the part life more predictable. If the life distribution is narrow, the data can be used to set up time- or use-based PMs to replace or service the part before it fails in the future.

Maintaining a part log is not as complicated or time-consuming as it may sound. On one of Agilent's most complex pieces of equipment—containing many thousands of parts—we placed fewer than 50 of these parts in the machine's part log.

Quality Checks

In an ideal manufacturing environment, no quality tests would be required; perfect quality would be assured by controlling all the inputs to the quality process, including the condition of all machine components. But until perfection is attained, product quality tests must still be carried out on machines used in the manufacturing process.

Agilent uses multiple types of quality checks to inspect machine performance. Some quality checks are made by operators on the actual product wafers coming out of a machine. These quality tests are not considered PMs; rather, they are regarded as part of the documented production operating procedure.

In some cases, quality checks cannot be performed on production wafers and a "quality check PM" must be scheduled. Agilent schedules these quality checks like any other PM, except they are most often performed by production operators.

Figure 5-9 shows a typical quality check PM.

Document #A-FCF1-0181 –1 Rev L
Dielectric Deposition Monitor Checklist

Operator _____ Date _____ Time _____

Machine Name _____ Chamber Number _____

1. Measure BEFORE particles on (2) Recycled wafers (BHRECYCLE). If these wafers are not available, substitute (2) PRIME wafers (P/N 4340-0393).

   The BEFORE count must be less than 25 on recycled wafers or less than 10 on PRIME wafers in order to proceed with this process monitor.

   Particle Measurement Tool: SURF6420-1

   Particle Measurement Recipe: DIELDEP.84

2. Record the wafer numbers and BEFORE particle counts in Section 6 below.

3. Place the monitor wafers in the machine and deposit dielectric on them using recipe DIELDEP.

4. Measure and record in Section 6 below the AFTER particle counts on both wafers.

   Particle Measurement Tool: SURF6420-1

   Particle Measurement Recipe: DIELDEP

5. Measure the deposition film thickness on each of the two wafers and record in Section 6 below.

   Thickness Measurement Tool: UV1250-1

6. Recorded Data:

Wafer #	Before Particles	After Particles	Thickness Avg	Std Dev
_____	_____	_____	_____	_____
_____	_____	_____	_____	_____

Enter the data into the SQCS quality control program. Follow the SQCS instructions if data are out of spec.

Place the "USED" wafers in the Recycling Box

Figure 5-9. A typical Agilent quality monitor checklist

STEP 2 MASTER CHECKLIST

When a TPM team's maintenance plan is organized and documented, the team is ready for a Step 2 final audit. Successfully completing all of the line items on the master checklist completes the Step 2 activities.

STEP 2 MASTER CHECKLIST: IDENTIFY COMPLETE MAINTENANCE PLANS
All maintenance plans have been identified, considering:
☐ Vendor-recommended maintenance plans
☐ Your actual experience with the machines, particularly areas of recurring failure
☐ Components with high failure rates are being maintained
All maintenance plans have been documented:
☐ Cleaning and inspection standards for operators
☐ Preventive maintenance plans for maintenance technicians
☐ Quality checks for operators and maintenance technicians
PM checklists are completed for each PM:
☐ All PM checklists are officially documented: ____ Checklists comply with the company's content and style guide ____ Checklists are controlled documents
☐ A master PM plan report is assembled for each machine: ____ The master plan includes all time-, use-, and condition-based PMs ____ No other system or method is employed to schedule or document PMs
Every PM checklist item that specifies an inspection provides criteria or specifications for an objective inspection
Every PM checklist item that specifies a part replacement has a specified part number and part description
Every checklist item references a written procedure (except generic machine procedures)
☐ Procedures can be an internal company document referenced by the document number, title, and section number
☐ Procedures can be vendor documents referenced by name, document number, and page number
Every critical part life is tracked by a dedicated PM or in a part log

STEP 2 INFRASTRUCTURE SUPPORT

In order for TPM teams to complete all the items on the Step 2 checklist, certain systems or infrastructures must be created within the organization, if they do not already exist. The TPM steering committee will need to see that the following systems are created for TPM teams completing Step 2:

1. A PM scheduling system capable of:
   - Time-based maintenance
   - Use-based maintenance
   - Condition-based maintenance
2. A part log system
3. A PM checklist system
4. A maintenance procedure documentation system
5. A spare-part management system
6. Changes in evaluation, ranking, and reward systems to support wanted behaviors and stop rewarding old unwanted behaviors; for example, reward contributors to the development of systematic maintenance planning.

STEP 2 DELIVERABLES

After completing Step 2, a TPM team should have a completely planned, scheduled, and documented technician preventive maintenance plan for their machines. This plan will not be perfect and will not prevent all machine breakdowns or losses, but it is the required foundation for improvements in later TPM steps.

STEP 2'S MINDSET CHANGE
Creating and using a maintenance plan effectively requires using detailed checklists to carry out scheduled maintenance work. It also requires adhering in a disciplined way to checklist criteria during PMs. This is different from allowing individual technicians to decide what any given PM should consist of and how it should be performed.

6

Step 3: Implement Maintenance Plans with Precision

This chapter describes how to implement TPM Step 3, including:

- Performing PMs on time and completely
- Executing maintenance work with precision
- Providing training for precision maintenance skills
- Providing precision documentation, including visual and text documentation
- Organizing the visual workplace, including tools and spare parts management
- Elevating knowledge, skills, and maintenance roles

6

STEP 3 GOALS

In Step 1, activities were performed to restore equipment to "new" condition by making all areas of the equipment clean and free of humanly detectable minor defects.

In Step 2, preventive maintenance (PM) plans were identified, organized, and scheduled.

The goal of TPM Step 3 is to achieve PM implementations consistent with the documented maintenance procedures developed in Step 2. This is achieved when "the machine does not know the difference" as to who is performing the maintenance work. This applies to all types of maintenance procedures, whether carried out by technicians or operators.

The result of Step 3 activities is a continually rising level of predictable, consistent, and desirable maintenance behavior known as precision maintenance. Precision maintenance is 100 percent complete maintenance carried out on time and executed precisely according to the prescribed procedures. As precision maintenance improves, so does equipment performance.

To achieve precision maintenance, maintenance technicians primarily require *tools* and *knowledge*. These must be systematically provided by factory management, as outlined in the remainder of this chapter. Without these, maintenance technicians cannot succeed. Success also requires maintenance technicians to carry out PMs with discipline and to pay attention to details in their work. This discipline must be achieved before more creative maintenance work—developing improved maintenance plans and machine designs—can be undertaken. If an organization cannot carry out its current maintenance plan with any reasonable degree of precision, improvements in the plan will not be effective and, in fact, may just create more chaos in the way that maintenance work is performed.

Unfortunately, this critical concept is overlooked in many TPM efforts. The focus of most improvement activities is on results, of course—in this case, improved machine and factory productivity. But new results come from new behaviors. TPM is a change program that changes behaviors first—people's changed behavior then achieves the desired results.

 Precision maintenance is 100 percent complete maintenance carried out on time and executed precisely according to the prescribed procedures. It is achieved when "the machine does not know the difference" as to who is performing the maintenance work.

ON TIME AND COMPLETE

Many companies struggle to complete PMs on time. In many cases this is due not to scheduling problems within the maintenance department, but to demands of the production department to continue operating the equipment. Few production managers are willing to stop the operation of a running machine to perform preventive maintenance. But to get out of the vicious cycle of breakdown maintenance, this is exactly what must be done. This TPM step involves a critical behavior change that managers must lead in their organization in order to change tech work from that of repairing broken machines to maintaining machines which always operate properly.

Agilent's IC fab management made this commitment by setting a policy that machines will not be run in production if they have an overdue PM. There is a window in which the PM can be performed—the longer the PM interval, the larger the window of opportunity—but if the PM remains incomplete after this interval, the machine cannot be run in production. We automated this by having our shop-floor control system check with the maintenance scheduling system before every production lot start. The machine is automatically locked down if a PM is overdue.

However, such automation is not necessary—only the management commitment that overdue PMs will not be permitted. In our case, we found that there are no overdue PMs simply because our management system will not permit them without severe consequences—shutting down the production equipment.

The next phase of precision maintenance is the thorough completion of every PM checklist. This means that every item must be completed as specified by its procedure. Items cannot be skipped over or left half completed. This is a clear expectation of all people performing maintenance. Compliance to cleaning and inspection standards carried out by operators is fairly easy to achieve, simply by implementing an ongoing cleaning and inspection audit system.[1] The same expectation is set for maintenance technicians performing PMs on equipment. Managers need not police this policy by observing or directly auditing tech PM behaviors, but they do need to reinforce it every way possible.

Frequently, the principal reason for an incomplete PM is the tremendous pressure placed on maintenance technicians to get the PM done as quickly as possible so the machine can be placed back into production service. This is the same reason that breakdown repairs are typically of the "bailing wire and chewing gum" variety—not because this is the level of the technician's capabilities, but because the tech is providing what the managers expect: minimum repair time for down equipment.

In Agilent's IC fab, each PM has a standard time established that is normal for its completion. Establishing standard PM times and tracking actual PM times provides a useful multipurpose tool:

- Standard PM times create a specific interval for maintenance technicians to complete a PM completely and correctly. This takes unnecessary pressure to take shortcuts in their PM work off technicians. The granted downtime also sets an expectation for technicians to perform their work thoroughly and at a reasonable pace.
- Actual PM times that are longer than the PM standard time may indicate problems carrying out the PM. The TPM team should review this situation. Is the standard PM time too short? Are there difficulties encountered in performing the PM? What improvements can be made?
- Consistently shorter PM times might indicate two different possibilities. Perhaps improvements have been made in the PM execution, and the standard time can be reduced. Or perhaps shortcuts are being taken to complete the PM, which may be inappropriate. Again, the TPM team should investigate the situation and make appropriate corrections.
- If a PM was completed once in an unusually short time, it may mean that the PM was simply "pencil whipped" to completion. This practice must be stopped.

---

[1] For more information on a cleaning and inspection audit system, see "Ongoing Cleaning and Inspection Audits" on page 105.

Providing technicians with the time they need to carry out the PM *completely* and *accurately* usually goes a long way toward eliminating incomplete PMs. Managers need to keep in mind that skipped maintenance items often show up later as machine failures.

In short, 100 percent completion of PM checklists is not an impossible task. However, to achieve it, maintenance managers must set this expectation for their equipment techs and then provide the required support and resources. Managers also need to provide positive reinforcement for compliance with this expectation.

In addition to requiring that PM checklists be completed, Agilent maintenance managers require techs to have written PM checklists with them at the equipment every time they are completing a PM.

PRECISION EXECUTION

Maintenance work that is carried out with precision has a single result: the machine does not know the difference among the people who carry out the maintenance work. In other words, to get the machine to behave consistently in production service, maintenance procedures must be performed consistently, even when being done by many different people. Ensuring that all maintenance people follow the same procedures when performing maintenance is not an easy task.

However, there are four basic tool sets that can be used to help a maintenance organization achieve precision maintenance execution:

1. Provide basic technical skill training for generic machine components and technologies
   - Deliver training with proficient methods
2. Provide training for specific machine skills for the different types of machines used in the factory
3. Provide training to develop failure prevention skills
4. Provide precision maintenance support tools:
   - Precision documentation
   - 1-Point Lessons
   - Time to learn
   - A "Partner and Practice" system
   - An organized workplace
   - Spare part and tool management

Basic Technical Skills

Maintenance personnel must have mastered the basic skills of their craft in order to carry out precision maintenance. Agilent technicians have all graduated from an accredited two-year electronics technical school or have the equivalent experience, such as military training. We expect the maintenance technicians that we hire to have this basic electronics training before they begin their jobs because our equipment contains so many electronic components. Techs without these skills could never support our equipment. Many Agilent operators have become maintenance technicians, but only after acquiring a two-year technical degree in electronics.

Agilent technicians must maintain an incredibly diverse array of complex equipment, mechanical as well as electronic. Achieving this goal requires them to make continual advancements in their equipment knowledge and maintenance skills—particularly those they did not acquire in technical school. Technical training continues for our maintenance people forever, encompassing about 10 percent of their entire career time.

Technicians must continually improve their basic technical skills in order to properly maintain equipment and prevent equipment failure. Machines are constructed of many types of common components, and technicians must have a basic command of each of these technologies. This can be achieved by setting up a series of "minicourses," most lasting about two hours. These courses can be set up and taught in-house or contracted out to professional technical trainers.

The following are some of the training subjects identified as useful in Agilent's IC fab. A company implementing TPM needs to inventory the types of components and technologies used in their factory and designate training courses that are appropriate for their own maintenance technicians.

- Fasteners, threads, and thread-locking devices
- Fluid transmission components
  - Tubing and hose
  - Valves
  - Seals—static and dynamic
  - Pumps
  - Fluid fittings:
    Pipe thread fittings/swivel pipe adaptors
    JIC (AN) fittings
    SAE O-ring boss fittings
    Swagelok tube fittings
    VCO fittings
    VCR fittings
    Ulta-Torr vacuum fittings
    Quick-connect fittings
    Flanged fittings
- Pneumatic system hardware and schematics
- Hydraulic system hardware and schematics
- Mechanical drawings

- Materials:
  - Metals
  - Plastics
  - Ceramics
- Electronics:
  - Basic electronics skills
  - Circuit boards and schematics
  - Cable assemblies
  - Connectors
  - Computers and PLCs
  - Motion controllers
- Electrical power transmission:
  - Volt/ohm meters and other basic electric instruments
  - Single-phase power
  - Three-phase power:
    Four-wire systems
    Five-wire systems
    Y connections
    Delta connections
- Motors:
  - A/C
  - D/C
  - Stepper
- Power transmission:
  - Bearings
  - Lubrication
  - Belts and pulleys
  - Chains and sprockets
  - Gears and gear boxes
  - Couplings—solid and flexible
  - Cams and cam followers
  - Keys and key ways
  - Splines
  - Press and slip fits
  - Snap rings and C-rings
  - Set screws and taper-locks
  - Clutches and brakes
- Adhesive, caulks, and sealants
- Compressors

- Agilent industry-specific tools:
  - Gas analyzers
  - Oscilloscopes
  - HVAC tools
  - Refrigeration systems
  - Vacuum systems
  - Chemical and gas systems
  - Cryogenics

Many of these subjects may seem trivial and unnecessary. Very common attitudes among maintenance technicians include, "I already know all that I need to know about these subjects," and "I will pick up all that I need to know about these subjects just by working with them on the job." Unfortunately, the lack of specific, detailed knowledge about many types of common machine components causes people to actually create minor abnormalities in machines rather than remove them. It is not uncommon for half of all the machine abnormalities found in a piece of equipment to be man-made rather than the result of machine deterioration. The more that technicians learn about basic machine components, the better they will be able to carry out precision maintenance.

Maintenance techs will not just "pick up" all of the details needed to master precision maintenance. If you don't provide a means for them to learn what they need to know, don't expect them to know it. Agilent electronic techs were not hired with a detailed knowledge of plumbing fittings— they acquired these skills through Agilent training.

Preparing Technical Minicourses

No two companies train their maintenance technicians using exactly the same set of training classes. Basic training must be customized to the company and industry that people work in. For example, Agilent teaches a course that covers the nine most common fluid fittings used in our equipment, as noted in the above list. However, were we in the refrigeration industry, the course would include flare fittings. Were we in the agricultural industry, it would include cam-lock fittings.

The best way for a company to develop appropriate courses in basic machine technologies is to take stock of the most common generic components used by their techs. Threaded fasteners are certainly a common component, but Agilent's course on threaded fasteners includes quite a bit of information about stainless steel fasteners. Many industries would not include this information because they use no stainless steel fasteners.

Once 20 to 30 common maintenance technologies have been identified, the construction of a course on each subject can begin. A good way to decide on the contents of each course is simply to observe the tasks that machine techs perform on these types of components. For example, Agilent equipment contains a large number of pipe threads. Our pipe threads are all assembled with Teflon tape to provide a leak-tight joint and ensure against contaminating the fluid inside the pipe system. Some obvious questions to answer in a class on pipe fittings include:

- How much Teflon tape should be placed on the pipe thread?
- Which way should the Teflon tape be wound?
- How many threads should the Teflon tape cover?
- When tightening a pipe joint, how much should the male pipe be inserted in order to have a properly assembled joint?

The course on fluid fittings included in Appendix A[2] is a sample of the kind of miniclass that teaches basic skills to maintenance technicians about machine components they frequently encounter. There are hundreds of different kinds of fluid fittings used in the world, and dozens of different kinds used throughout Agilent's facility. But there are only about nine types of fluid fittings that are found on Agilent's factory floor equipment. Every maintenance technician is expected to know how these nine fittings work, how to inspect them, and how to properly use them.

Note that there are exercises throughout the course that require students to actually use the components they are learning about. They will not learn what they need to know about maintaining these components just by reading about them or by listening to someone discuss them. The courses are designed to follow four simple guidelines for teaching and learning:

- *Acquire the technology.* You can't teach what you don't understand, whether it is procedures for tightening bolts or lubricating bearings.
- *Break the material down into small pieces.* Two-hour training sessions are the norm. People require time to acquire new knowledge and skill. Overloading them is a sure way to achieve poor comprehension.
- *Learn by doing.* Require the students to actually work with the machine components they are learning about. People also learn best when studying a problem or task that is important to them.
- *Provide guided practice.* The trainer should be certified in the material being taught and should have a well-organized class agenda that allows people to practice with the material under the trainer's guidance.

---

[2]The abbreviated course in Appendix A teaches techs about four types of fluid fittings. It is included in this book as an example of the kind of knowledge that maintenance technicians should obtain about basic machine components.

Most manufacturing companies would want their maintenance technicians to take a course on threaded fasteners. Such a course could be created by observing how technicians use various threaded fasteners in their work, the types of fasteners that are stocked in spare parts, and the types of tools that are kept in technicians' toolboxes. This was done at Agilent, and the following line of questioning illustrates the process that created our minicourse on fasteners.

Agilent toolboxes in the IC fab include both Phillips and Pozidrive screwdrivers. They appear to be very similar, as shown in Figure 6-1: the top driver is a #2 Pozidrive, while the lower driver is a #2 Phillips.

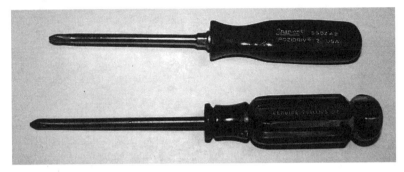

Figure 6-1. These screwdrivers appear similar

- What is the difference between a Phillips and a Pozidrive screwdriver? Are they interchangeable? If not, how do you know when to use one or the other?
- On Phillips head screws, when do you use a #1 Phillips? a #2 Phillips? a #3 Phillips?

Different kinds of hex head bolts are stocked in Agilent's spare parts bins (Figure 6-2).

- What do the different marks on the top of these hex head bolts mean?

No Lines   3 Lines   5 Lines

Figure 6-2. Hex head bolts with different markings on their heads

Socket head bolts are also stocked but have no such markings on their heads (Figure 6-3).

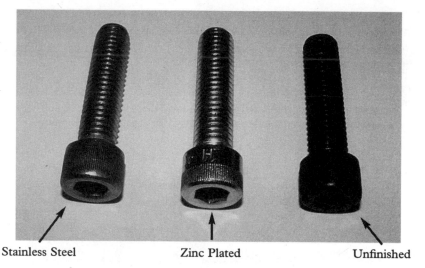

Stainless Steel   Zinc Plated   Unfinished

Figure 6-3. Different kinds of socket head bolts

- How do you tell the difference among a variety of socket head bolts?
- How do socket head bolts compare in strength to different kinds of hex head bolts?

Different kinds of lock washers are stocked in Agilent's spare parts bins (Figure 6-4).

- When do you use a split lock washer versus a star lock washer?
- What is the difference between an external star and an internal star lock washer?
- Why do some split lock washers look different from others?

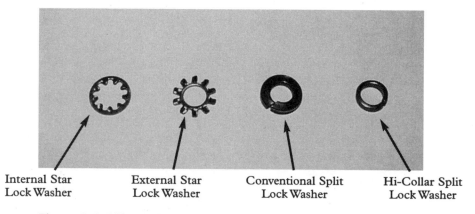

Internal Star   External Star   Conventional Split   Hi-Collar Split
Lock Washer   Lock Washer   Lock Washer   Lock Washer

Figure 6-4. Different kinds of lock washers

Other questions our technicians asked about threaded fasteners and their use included:

- What do I have to know in order to specify a bolt completely?
- For a standard nut-and-bolt assembly, what is the weak link—the bolt shank, the bolt thread, or the nut?
- Which are stronger—coarse or fine threads?
- How do I know how much to tighten a bolt?

These questions led to the development of a maintenance technician course on threaded fasteners, which taught Agilent technicians the answers to these and other questions. (For the curious reader, abbreviated answers to the above questions are provided in Appendix B.)

Training Guidelines

Once training requirements have been defined, maintenance training needs to be implemented in an effective manner. Agilent uses the following guidelines for operator and technician training processes.

- People can only perform production or maintenance procedures that they are certified to carry out.
- All certification requirements are documented within the training material.
- A timetable exists for each person to complete the required training certifications.
- Training is delivered by a certified trainer.
- Multiple techniques are used to deliver training, depending on the material, and often are tailored to how individual students learn best.
- Certification is completed by a designated certifier who is not necessarily the certified trainer.
- The goal of certification is demonstrated competence.
- Annual recertification is required in some disciplines such as safety, wafer handling, and ergonomics.
- Quarterly random audits are carried out. These are prescribed in detail in the training material. Auditors may randomly ask operators and technicians the following questions.
  - What are you doing?
  - Are you certified to do it?
  - Demonstrate the correct procedure for your current activity.
  - Show me where you would find the document describing this procedure.

Specific Machine Skills

The equipment found in Agilent's IC fab is quite complex. The technicians who maintain it must understand in great detail exactly how these machines operate.

This requires many hours of study, usually at the machine vendor's school, on each piece of equipment. Agilent technicians cannot perform any maintenance activities on complex machines until they are certified to do so. Such certification comes from training classes carried out on the specific machine. The more complex the machine, the more training at the vendor school our techs receive. It is not uncommon for Agilent techs to receive two to three weeks of specific machine training (in one-week intervals spread out over a year) on one piece of our IC fab equipment.

Failure Prevention Skills

In order to prevent equipment failures, technicians must master more than machine maintenance skills. Even maintenance technicians with years of training in machine components and experience in maintaining machines cannot be expected to know how to prevent equipment from breaking down. Preventing equipment failure is a new technology for most maintenance people. They must learn the tools and methods of failure prevention. These methods—described in Step 4 of this book—are taught and practiced in Agilent training sessions.

Precision Maintenance Support Tools

As stated above, precision maintenance support tools include:

- Precision documentation
- 1-Point Lessons
- Time to learn
- A "Partner and Practice" system
- An organized workplace
- Spare part and tool management

Precision Documentation

Maintenance people cannot carry out their work with precision if they are not supplied with the necessary resources. These include, of course, the necessary hand tools, instruments, and spare parts. Yet the most important tool—and the one most often lacking in many organizations—is useable documentation of maintenance procedures. Some maintenance procedures are simple and generic enough that documentation is not required, but any manufacturing facility probably contains a number of complex machines, the maintenance procedures for which are far from generic or obvious. Simply stated, precision work cannot be completed without properly documented procedures.

Successful maintenance usually depends more on tools and knowledge than on skill or craft. Most machine maintenance is neither rocket science nor art. Rather, it is a set of precise procedures that require the proper tools, parts, and knowledge.

A maintenance technician supplied with the correct procedure in a useable form, and all the right tools and parts, has a good chance of doing the job precisely.

 In a sense, the probability that a maintenance procedure will be done precisely as it should be done is often 80 percent determined before the maintenance technician even begins the work.

Unfortunately, most maintenance documents are not very useable. Many consist of pages of text with a few illustrations that are ambiguous and nearly impossible to understand. Most technicians will not even attempt to use such documentation—it will simply languish on a shelf somewhere until someone is assigned the task of "updating" it in hopes of maintaining compliance to documented procedures. Precision maintenance requires documented procedures that are easily useable and totally unambiguous, and which contain all the detailed information required for any maintenance tech to complete the task and obtain the proper end result.

Agilent has experimented with different documentation systems, including multimedia documents available to the technician at the machine on the factory floor. However, the simplest system turned out to be the best one for us: we simply describe the maintenance procedures with pictures instead of paragraphs of text. The pictures are easier and faster to create than written instructions, so this type of document is actually simpler and cheaper to produce than typical text documentation. Usually a few minor notes added to the photos provide all the information a maintenance tech needs to perform precision work. These documents can be produced on paper or placed on-line, whichever works best for an organization.

The following pages were taken from a document that explains how to rebuild the wafer-handling robot shown in Figure 6-5. Once the robot's assembly procedure was well understood, this 300-picture document was produced in less than a week.

It is also a good idea, after creating such a document, to have someone not involved in its creation "practice" using it. Next time the work is required, have this person use the document to perform the work. This practice session should reveal any weaknesses in the document, including ambiguities and omissions. After only one practice session, the visual document is usually perfected to the point where any technician can perform this work—*with precision*—using generic craft skills and the visual document.

Figure 6-5 shows a rebuilt wafer-handling robot assembly. Figure 6-6 shows the robot disassembled and ready to be rebuilt. The rebuilding procedure for this subassembly was originally an all-text document that was replaced by a visual assembly document.

Figure 6-5. A completely rebuilt wafer-handling robot. This machine requires precision assembly, which cannot be done without precision documentation.

Figure 6-6. A disassembled robot ready to be rebuilt

The following sample from the text document was so ambiguous and difficult to understand that it was never actually used, even though it was factually accurate. Most techs just ignored it and assembled the wafer handler from memory, which caused many assembly errors.

Read the following text instructions and try to comprehend what the text is telling you to do. Then look at the visual document that replaced this text—shown in Figures 6-7 through 6-9—and see how it is inherently more understandable.

Mounting the Robot Base on the Wafer-Handler Rebuild Cart (Text Instruction)

1. Stand on the long side of the robot rebuild cart with the cart handle to your left. Rotate the cart platform so it faces up—that is, the bar that prevents total platform rotation is on the bottom of the platform and is on your right side.
2. Set the robot base in the center of the platform, right side up as it would be in the machine. Two pins protrude through the robot base preventing it from sitting flat on the cart platform unless the robot base is in one of two positions, the pins being on the right and left sides of the cart's long axis.
3. Only one of these two positions is correct. Choose the one that, when the platform is flipped bottom up, has the dowel pin hole nearest the center bore of the robot base on your right side.
4. Mount the robot base to the cart platform using six 3/8-16 × 1″ socket head bolts. With the cart handle on your left side, the side of the cart closest to you is hereafter referred to as the NEAR SIDE of the robot, and the side of the cart farthest away from you is referred to as the FAR SIDE of the cart.

Mounting the Robot Base on the Wafer-Handler Rebuild Cart (Visual Instruction)

Figure 6-7. References from assembler's point of view

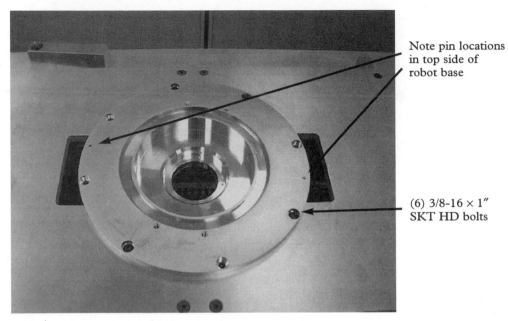

Note pin locations in top side of robot base

(6) 3/8-16 × 1″ SKT HD bolts

Figure 6-8. Locate the robot base on the service cart

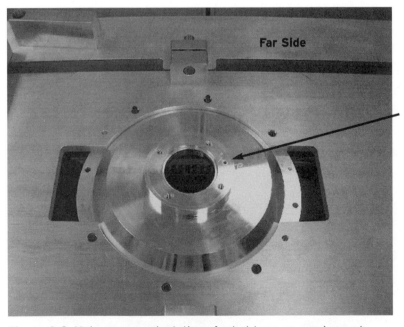

Far Side

Note dowel pin hole location in bottom side of robot base

Figure 6-9. Note proper orientation of robot base on service cart

Documentation that is created but not used is a waste of resources. Useable documentation, on the other hand, helps technicians successfully carry out precision maintenance work. Visual documentation is often far more useful in describing maintenance procedures than text documentation.

More examples of visual rebuild procedures follow, showing the steps required to assemble just one small bearing joint in the entire assembly. The old adage, "A picture is worth a thousand words," rings true when it comes to creating useful maintenance documents for equipment technicians. Imagine using a text document to describe the detail that is provided to the assembler in these simple photographs (Figures 6-10 through 6-14).

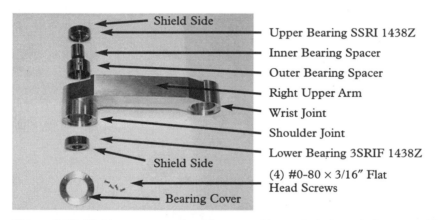

Figure 6-10. Right upper arm shoulder assembly—exploded—ready for assembly

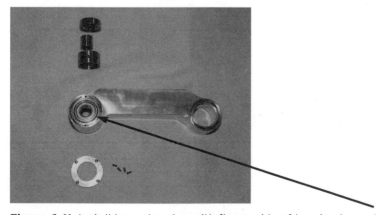

Figure 6-11. Install lower bearing with flange side of bearing toward outside

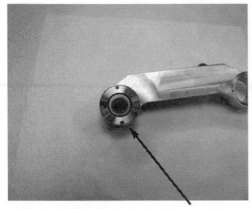

Figure 6-12. Install lower bearing cover.

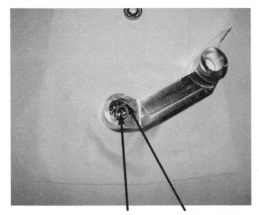

Figure 6-13. Install inner and outer bearing spacers

Figure 6-14. Install upper bearing with shield side outward

Following is another example of an Agilent maintenance procedure that was upgraded from text to a visual document. Note the clarity—completely lacking in the ambiguous text—achieved with these photographs (Figures 6-15 through 6-17).

1. Mount two optical switches onto the P132/P133 bracket using #4-40 × 1/4″ SKT HD screws with lock washers. The heads of the screws should be on the switch side of the bracket. Mount the optical switches with the emitters away from the bracket. (The emitter side has the chamfer.)

2. Mount two optical switches onto the P130/P131 bracket using #4-40 × 1/4″ SKT HD screws with lock washers. The emitters should be facing the opposite direction on this bracket from the P132/P133 bracket; otherwise the emitter on one sensor set will interfere with the other sensor's detector.

> This last sentence is so ambiguous that it has caused many misassembly problems.

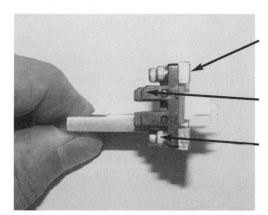

Bracket P132/P133

Emitter positions (chamfered edges)

#4-40 × 1/4″ SKT HD screw w/LW

Figure 6-15. Assemble the P132/P133 optical switch assembly

Emitter positions (chamfered edges)

#4-40 × 1/4″ SKT HD screw w/LW

Bracket P130/P131

Figure 6-16. Assemble the P130/P131 optical switch assembly

FYI—The optical switches can be attached to the robot assembly in only one orientation. Their emitters will be positioned at opposite ends, preventing electrical interference between them.

Figure 6-17. Optical switches assembled on robot

1-Point Lessons

It would seem that classroom training and hands-on experience would be sufficient for maintenance people whose goal is precision maintenance. Yet, despite extensive classroom training, half of all the equipment abnormalities found in Agilent's equipment are still man-made. The overwhelming root cause for these abnormalities is lack of knowledge by the maintenance performer. Precision maintenance requires mastering thousands of small details—each one, separately, easy to understand.

But even the small details are important and, like all important things, must be handled just right. For example, if a 10-32 stainless steel thread is tightened into an aluminum block, which torque spec for the fastener should be used—the one for stainless steel or the one for aluminum? The difference is significant—32 inch-pounds versus 19 inch-pounds. The wrong choice could cause the fastener to be either over- or undertightened—either one being a minor defect that could significantly affect the machine.

If a radial bearing is dropped into a bore and must carry a thrust load, which way does it get inserted, face up or face down? When reaching for a #10 lock washer from open stock, when do I use the split lock washer? The internal star lock washer? The external star lock washer?

The lack of knowledge of thousands of small details like these causes man-made defects in equipment. These defects cause intermittent machine performance problems and early component failures, ultimately leading to lost machine productivity.

Any technician, no matter how experienced, can always learn something new about machine components and maintenance procedures. A 1-Point Lesson system is a good way for maintenance techs—and equipment operators—to continually learn new details that help them improve their precision maintenance practices. 1-Point Lessons fill a need not completely satisfied by traditional classroom instruction—the need to continually learn thousands of small details. Most of these lessons take only a few minutes to learn and can be prepared for distribution to others in about 10 minutes. People cannot absorb a great number of simple details lumped together in a weeklong classroom situation—they need to learn them one at a time, just one or two every day. This is exactly what a 1-Point Lesson system provides.

1-Point Lessons should be written by everyone—any time that someone learns something new about their job or equipment. In this way, individual learning becomes organizational learning. If every operator and technician could each learn one new detail to improve their performance every working day, the organizational learning rate—and ultimately, the machine performance rate—would continually improve at a rate far beyond what most companies achieve. A good rule of thumb is that "everyone should learn at least one new lesson every single day."

Sharing such a large number of 1-Point Lessons is not as difficult to achieve as it might seem. Suppose in a 100-person organization, each person wrote only one 1-Point Lesson a week. This amounts to about 5,000 lessons per year. Even if only one fourth of these lessons—about 1,250—were relevant to any one person, that person would receive 25 lessons each week, or about 5 per average working day. Once a 1-Point Lesson system is introduced into an organization, and people learn to use the lessons to capture small important details, the learning rate of the organization can be increased tremendously.[3]

Time to Learn

One of the important lessons learned repeatedly during Agilent's TPM activities is that people need *time* to learn how to solve problems and carry out their work with improved precision. Typically, there is such intense pressure just to get the basic job done that people don't have time to learn something new that might prevent future difficulties or reduce future workloads. Improvements in both human and machine performance come from new learning and the development of new skills; however, this will occur only if people are granted time by their managers to engage in learning activities.

For instance, some techs at Agilent were engaged in repairing a machine failure that repeated about every 10 days—a roller chain in a vacuum chamber

---

[3]For more information on 1-Point Lessons, see "1-Point Lessons" on page 78.

whose roller pins "locked up" and refused to roll around the chain sprocket. The techs had replaced the bad chain five times already; but every 10 days, the replacement chain would fail again. Because the factory floor was running at full capacity, the techs were under tremendous pressure to replace the broken chain and restore the machine to production as quickly as possible. Managers believed pushing for this "fast repair" would maximize wafer production through the IC fab.

Finally, a maintenance manager realized that the repeating chain failure—not the time taken by the technicians to replace the chain—was the real culprit in slowing wafer production. So the manager gave his techs a new directive: replace the broken chain, but keep the machine down as long as necessary to discover the reason for the repeating failure. The techs discovered that a recently installed part was improperly machined and was causing the chain to be misaligned. The strain of the misalignment was bending the chain roller pins. By correcting the improperly machined part, the chain never failed again—even after five more years of service.

The manager who gave his technicians "time to learn" by telling them to keep the machine down longer than the time required to make repairs actually helped increase wafer production that month. Unfortunately, allowing time to learn goes against common wisdom, which dictates that maximizing production means keeping machines up and running for as many hours a day as possible.

If people keep behaving the same way that they always have toward machines, the machines will continue to behave the same way as well. Someone once said that one definition of insanity is doing the same thing over and over the same way and expecting to get a different result.

Managers need to give their people time to learn, time to develop new skills, and time to analyze the root cause of failures. This "extra time" is not really "extra" at all—neither for machines or people. Rather, it is time invested in reducing peoples' workloads and machine failures. The rewards of such an investment can accrue very quickly.

A "Partner and Practice" System

Using all of the tools described so far is not enough to achieve true precision maintenance when different people maintain the same equipment. People must actually work together to discover the differences in their work procedures, which must then be resolved and the solutions documented.

A simple way of accomplishing this is to bring in a "guest tech" once a month on each shift in each work area to share a scheduled PM. For example, someone working on a Monday day shift could visit the Thursday night shift to help with a two-hour PM. If every tech made a visit to another shift's team just once a month to complete one PM, the interactions among techs performing the same maintenance work on the same machines would be considerable. This interaction would rapidly reveal the differences in their work and allow the documented procedures to be improved to eliminate these differences.

Many organizations send representatives to monthly meetings to talk about maintenance procedures in hopes of resolving differences. Some progress might thus be made, but those gains pale in comparison to the results achieved by people actually performing the work as teammates. This is when true precision maintenance finally commences.

An Organized Workplace

One way to improve the precision of all production and maintenance work on the factory floor is to provide people with a highly organized work environment, where much of their work is visually controlled. The visual workplace is ideally a zero-defect, zero-abnormality workplace.

Figure 6-18 is an overview of Agilent's visual workplace.

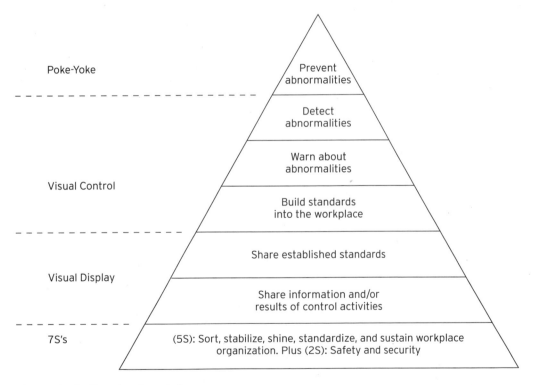

Figure 6-18. The visual workplace

Workplace organization is developed on a 7S foundation—5S workplace organization plus safety and security—as follows.

1. *Sort.* Distinguish between what is needed and not needed.
2. *Stabilize.* Provide a place for everything and maintain everything in its place.
   - Determine the best location for all identified needed items.
   - Determine how many of each item will be stored in each given location.
   - Make it easy for anyone to find, use, and return these items.
3. *Shine.*
   - Keep the workplace clean. Eliminate dirt, dust, and other foreign matter.
   - Adopt cleaning as a form of inspection. This exposes abnormal conditions and corrects prefailure conditions by identifying problems while they are still minor.
   - Integrate cleaning into everyday maintenance tasks. This adds value to equipment and pride in the work area.
4. *Standardize.* Maintain and monitor the first 3S's.
   - Standardize everything and make standards visible so all abnormalities can be easily and immediately recognized.
   - Devise methods to maintain adherence to the desired state and prevent deviations from standards to:
     - Prevent accumulation of material in unwanted places
     - Ensure everything is returned to its own place
     - Maintain cleanliness standards
5. *Sustain.* Stick to the rules scrupulously. Make sure:
   - Correct procedures become habit.
   - All employees have been properly trained.
   - Employee work habits have changed.
   - The managers of the organization are deeply committed to implementing and maintaining these first 5S practices.

In addition to 5S workplace organization, Agilent has added two more subjects to the foundation of our organized workplace.

6. *Safety.* Make sure every process and activity that is carried out on the factory floor is done safely.
   - Safety must be a proactive effort. The design of any new activity must be founded upon safe practices
   - Everyone must adhere to all safety standards at all times. Continual audits should be set in place to assure compliance with all safety rules.
7. *Security.* Make the work environment secure in several areas:
   - Provide security for the company's intellectual property.
   - Provide security for customers. Put systems in place that assure the reliable delivery of their orders:
     - Software and recipe backups
     - Factory fire and loss security measures

- Provide job security for employees. The company's success depends on the manufacturing processes always being implemented reliably and precisely:
  - Deviations from process standards are not allowed
  - Systematic improvements in process standards are encouraged

Creating a Visual Workplace The first step in creating a visual workplace is 5S workplace organization. A visual workplace is capable of providing the following benefits:

- There is nothing extra or unneeded.
- Storage areas are clearly distinguished.
- There is a place for everything and everything is in its place.
- The workplace is kept immaculately clean.
- Items, information, schedules, and processes are recognizable at a glance.
- It is easy to distinguish immediately between what is normal and what is not.
- Paperwork is simplified and minimized.
- Waste and abnormalities are immediately recognizable to anyone.
- The flow of product, deviations from standards, and everything else that exists or occurs in the workplace is readily apparent at a glance.
- Standard procedures are easily understood.
- Quality is increased.
- Productivity is enhanced.

Following are some examples of 5S principles that have been implemented in Agilent's IC fab.

Production wafer boxes are color coded as follows:

Wafer Box	Color Purpose
Yellow w/Red Lid	P-1 Priority lot
Yellow	P-2 Priority lot
Blue	Production lot
Red	Monitor wafers shared throughout the IC fab
White	Experiment lots
Green	Recycle and reuse wafers
Brown	Setup and calibration wafers
Violet	Dummy wafers
Orange	Recycled wafers
Black	Wafers being transported to test
Brown and White	Test scrap

Staging nests for lots moving into equipment are color coded. (A nest is a colored plastic holder which holds a box of wafers.)

Nest	Color Purpose
Green	Lot in process
Yellow	Lot in staging
Red	Quality check wafers

Figures 6-19 through 6-21 show some of the results of Agilent's visual factory floor.

Figure 6-19. Everything has a place and everything is in its place

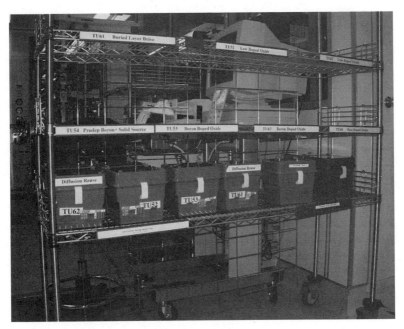

Figure 6-20. WIP racks are organized and labeled. Production material is color coded

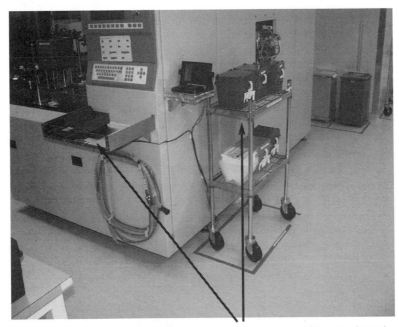

Figure 6-21. Production lots lined up to go into a machine are staged on color-coded nests at the machine's input station

The 5S process, like all Agilent IC fab processes, is audited for compliance and to continually improve workplace organization processes.

Spare Part and Tool Management

A spare-part management system is one of the most important systems in any manufacturing operation that supports a world-class maintenance program. No technician, no matter how skilled, can maintain good machine operation if the proper parts are not available to perform scheduled maintenance. Agilent has two kinds of spare part systems—open stock and controlled stock.

Open Part Stock Open stock parts are small, common, generic parts, including such items as nuts and bolts, electronic components, and fluid fittings. These parts are kept either on the factory floor or in technician service areas. Either way, technicians have direct access to them and can remove them as needed with no requisition process. Figures 6-22 through 6-24 show some of these part-storage systems. The parts are organized by part description, size, and part number. Cross-referencing databases for locating any part with just one piece of information are available at the part storage bin. Technicians should not waste their valuable time searching for an inexpensive generic part when they should be servicing a machine.

Open stock parts are sometimes owned by Agilent's supply vendors. They stock these parts in our factory in a highly organized fashion, inventory them regularly, replace what was used, and bill us for the replacements. This system keeps our overhead on maintaining some stocks of small generic parts to an absolute minimum.

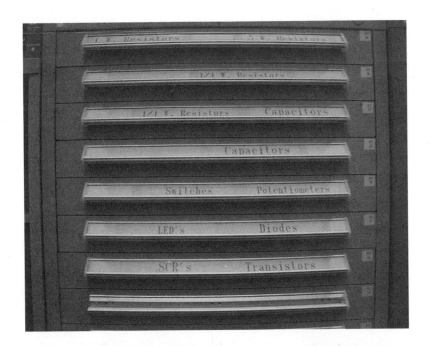

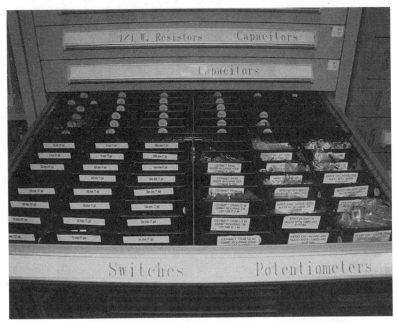

Figure 6-22. The assortment of open stock parts shown here is located in the maintenance technician service area, which is off the factory floor. None of these parts is scheduled to be used for PMs; these are general service and repair parts, and those used to support design improvement projects.

STAINLESS STEEL FITTINGS

Figure 6-23. The open stock parts shown here are located directly on the factory floor. They contain nuts and bolts and pneumatic fittings, which are used for both scheduled PMs and to make equipment repairs.

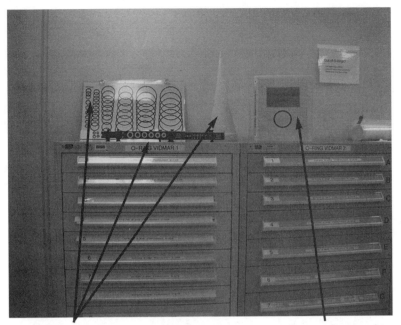

Sizing tools   Cross-reference data

Figure 6-24. The O-ring supply source shown here is open stock located on the factory floor. Parts are sorted by size, material, and part number, so the right part can be easily located by any number of means.

Controlled Part Stock Controlled stocks are parts that are available to technicians only on a requisition basis. These parts are owned and inventoried by Agilent, and the inventory quantity on hand is always reduced by the actual count of parts removed by technicians as they consume them. Figure 6-25 shows one of our spare part stock rooms and the requisition computer used to remove parts from the stock room. All controlled stock parts are purchased by Agilent buyers. Each part has a minimum "on-hand" quantity and an order quantity. When the inventory reaches the minimum stock number, the replacement order for the specified number of parts is placed automatically through Agilent's purchasing department.

Figure 6-25. Controlled parts are kept in a stockroom to improve the accuracy of inventory records. Technicians have direct access to the stockroom at all hours of the day or night, seven days a week. Parts removed from this location are required to be "requisitioned" on the inventory control computer shown at the front desk.

There are generally three levels of part-management routines—"A," "B," and "C" parts.

"A" parts are expensive. Our buyers spend most of their time managing our on-hand quantities of these parts. They are ordered frequently and in small numbers. We plan for their use and attempt to receive them on a just-in-time basis. This keeps our inventory of these expensive parts to a minimum while assuring that we always have what we need.

"B" parts are not necessarily expensive, but they are large and bulky. Because of their size, these parts cannot be kept in local stock rooms and so are stored in more remote warehouse locations. Storing and retrieving these parts when needed is a costly process; therefore, they receive a fair amount of buyer attention, although not as much as the more expensive "A" parts. They have minimum on-hand re-order points and replacement order quantities defined, as do all controlled parts.

"C" parts are small, inexpensive parts that make up approximately 80 percent of the parts we stock. We do not want to consume undue purchasing resources on parts like these, so we buy a six-month inventory at a time. The inventory cost is not excessive, and we spend little effort managing these parts, but still they are controlled and inventoried.

Two basic areas of our spare-part management system are constantly being addressed. First, we do not wish to have more inventory than we need, because carrying inventory is expensive. We review our usage history for every part twice a year and, based on that history and our vendor's lead time for replacement, determine whether or not we are carrying too much inventory. We are constantly seeking to reduce the cost of carrying spare parts. As TPM activities extend the life of parts and reduce failure rates, this semi-annual review allows us to reduce inventory levels.

The second spare-part management problem is lacking a part that a technician needs for a machine PM or repair. We regard any such part shortage as a part-system failure and treat it similarly to a machine failure: a failure analysis is performed and the problem is eradicated, no matter what its root cause. For this effort, we usually use our generic problem-solving tool, the PDCA.[4] The answer is not always to increase inventories, as might be suspected. Our management and order systems work fairly well when properly used. Most of our shortages are produced by three types of problems:

1. Parts are mistreated on the factory floor because of poor maintenance practices, which are caused by lack of knowledge or skill. This results in a sudden spike in parts usage. Improved maintenance procedures and technician training correct this type of problem.
2. Our actual inventory does not match our accounting inventory. This is caused by technicians removing parts from stock without going through the proper requisition procedure. In order to improve our inventory accuracy, a somewhat cumbersome requisition procedure was replaced with a simpler one.

---

[4]PDCA is described on page 201.

3. Our re-order points or order quantities need to be adjusted. Technicians are the users of our spare parts, and the system that replaces them; so it is up to them to improve these numbers in our system and thus improve its performance. Technicians have to take an active role in continuously improving our spare-part management system, just as they must continually improve the maintenance plans that they carry out on machines.

Maintenance Tools Supporting a factory full of complex equipment requires many kinds of tools. Standard technicians' toolboxes are, of course, required. Special service tools appropriate for the types of equipment on the factory floor are also necessary. In Agilent's situation—with so many vacuum chambers— many specialized vacuum tools are required, such as a helium leak detector and a residual gas analyzer.

Simple machine maintenance supplies are also required. Agilent technicians often found their maintenance tasks difficult to carry out for want of some very simple supplies like threadlocker, gasket cement, or glue. To prevent lost time searching for these items—some of which we did not even stock—the technicians prepared several supply boxes of these materials, each kept in a strategic location. This system keeps the supplies handy no matter where the technicians find themselves working. An inventory of the boxes' contents, including stock numbers, is kept in each box. If an item is missing or empty, a replacement can easily be pulled from the stockroom. Figures 6-26 through 6-29 show this type of technician supply and tool system.

Figure 6-26. A typical maintenance kit contains many odds and ends, which technicians need in their daily work. Numerous kits are kept in strategic locations.

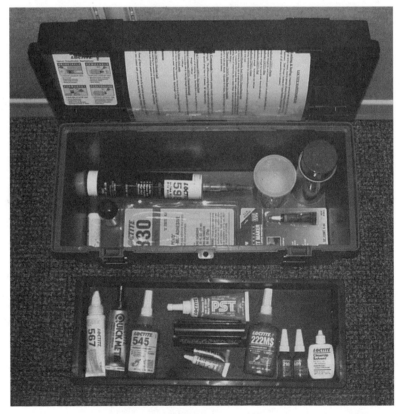

Figure 6-27. Instructions for use and a complete inventory identify all of the material kept in the maintenance kit. Items that have been used up can easily be identified and replaced from the stockroom.

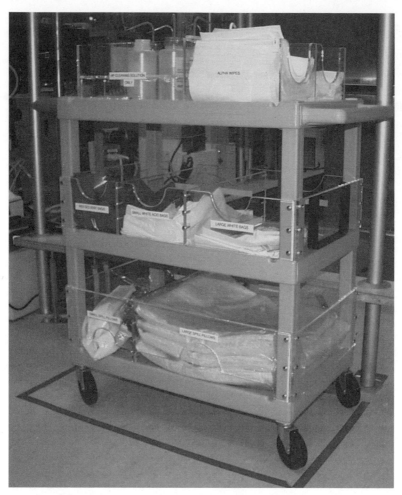

Figure 6-28. Cleaning supplies for daily contamination control activities in the clean room are organized and visually controlled according to the 5S process

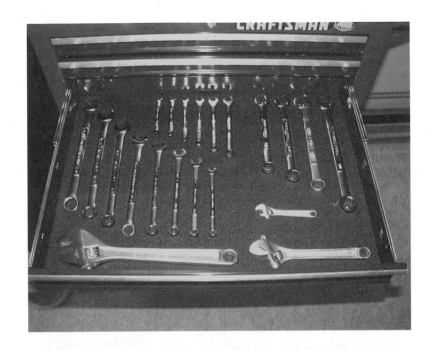

Figure 6-29. Maintenance tools are organized according to the 5S rules used to create a visual workplace for production material. Note how easy it is to spot the missing wrench in the toolbox in the top drawer. Before this improvement, techs found it difficult to find their maintenance supplies, as in the bottom drawer.

ELEVATING KNOWLEDGE, SKILLS, AND MAINTENANCE ROLES

To move to the highest level of precision maintenance, the knowledge, skill, and maintenance roles of all techs, operators, and engineers must be elevated to new levels.

Most Step 3 activities describe tools and techniques used by maintenance technicians to improve equipment performance with elevated maintenance routines. But the other TPM team members must also continuously elevate their knowledge, skills, and roles in improving machine performance.

Elevating the Operator's Role

Operators can continually advance their role in machine maintenance by continually *improving and adding to the operator cleaning and inspection standard.* The creation and use of C&I standards has already been discussed, and so have the audits performed on equipment to measure compliance with those standards. Audit teams can help increase operator involvement by pointing to ways that operators can expand their cleaning and inspection role on equipment. Members of TPM action teams can also create more 1-Point Lessons, training classes, special tools, or machine changes to expand and improve the operator cleaning and inspection role. This expansion process can continue indefinitely.

Operator participation in maintenance activities is only limited by their knowledge, their skill, and the role assigned to them by management. Most of the work done by people in any craft—whether that be plumbing or electrical—is accomplished by tools and knowledge. Add a little practice, and people can perform activities they never before thought possible.

Operators can continually take over low-level maintenance tasks from maintenance technicians. All a company need do is establish the level of the operators' maintenance role, and then provide them with the required tools, knowledge, and practice. This elevation in the operators' maintenance role frees technicians from spending time on these activities and allows them to pursue elevated maintenance practices themselves.

Operators can also continually improve on their own manual work. Agilent's goal is that 70 percent of all operators are as good at performing their routine jobs as our best operator, and the other 30 percent are not too far below this performance level.

Operators always have line-stop capability: They can stop the flow of product through the factory floor any time they find any abnormality in equipment or in a production lot. This is not only a right but a responsibility. Operators are our first line of defense against quality failures, and they should never process material through production equipment if they have any knowledge that something is amiss.

Operators' roles can be elevated to whatever degree a company deems appropriate, and not just in the area of expanded maintenance. At Agilent, our vision for future operator roles includes more participation in quality control of our manufacturing processes, identifying line limiters, capacity planning, and developing new line layouts and operating systems. This is complicated work, currently taken on mostly by process engineers and process technicians. But production operators could contribute a great deal to these tasks if provided with the right knowledge and opportunity.

Once a company has its own vision of operators' roles, it must take stock of its current position and plan a roadmap to achieve that vision.[5]

Elevating the Engineer's Role

A complete TPM program will also include activities for elevating the engineers' role in improving equipment and factory productivity. Of course, engineers on the TPM teams participate in all team activities, including setting machine standards, designating and improving maintenance plans, and helping the team with failure prevention and quality control.

But engineers can also elevate their role in preparing future manufacturing processes or equipment—a process often referred to as early equipment management. This involves the following activities:

- Including all improvements previously made in the design of a machine on a newly purchased machine of the same type before it is set on the factory floor.
- Including operators and technicians in early design reviews of new manufacturing processes, products, or equipment. This helps engineers enhance operability, maintainability, and safety in their new designs.
- Engineers can always improve their own technical knowledge in their own craft. Agilent IC engineers must continually stay current with the latest technology developments in the IC industry. This industry changes very quickly, and most Agilent engineers spend a large amount of their time studying to master the latest technology advances in their field.

STEP 3 MASTER CHECKLIST

When a TPM team has achieved PM implementations consistent with their documented maintenance procedures, the team is ready for a Step 3 final audit. Successfully completing all of the line items on the master checklist completes Step 3 activities.

---

[5] This elevation in operators' roles is part of TPM and is usually described in the higher-level steps of what is traditionally referred to as TPM autonomous maintenance.

STEP 3 MASTER CHECKLIST: IMPLEMENT MAINTENANCE PLANS WITH PRECISION

All PMs are completed on time

☐ A machine is not allowed to operate in production if the PM is overdue

All PM checklist items are completed when the PM is done

☐ No PM checklist items are skipped

☐ No extra material is left over from PM part kits

☐ PMs are completed in the standard time allocated, no matter who performs the task
- Standard times are set for each PM
- The actual time taken for each PM is tracked
- Deviations and their reasons are reported
- The reasons to improve consistent PM performance are acted on

Activities have been undertaken to improve the precision of PM executions by all the techs—the goal being that "the machine does not know the difference." These include:

☐ Techs come in off-shift to carry out PMs with other techs to synchronize their techniques

☐ Visual and written maintenance procedures have been improved or clarified

☐ The knowledge of "hundreds of minor equipment maintenance details" has been captured in PM checklists, PM procedures, and 1-Point Lessons

☐ All tools and spare parts have a place and are in their place, accessible by every tech

Evidence of successful precision is apparent:

☐ Audits of all new behaviors initiated in this step indicate that they have become normal

☐ The machine metrics to be improved by the team are being measured, and trend charts are posted on the Activity Board

Precision maintenance improvement activities have become part of the maintenance culture, so that continual improvement in precision will advance forever, despite the completion of this step

STEP 3 INFRASTRUCTURE SUPPORT

In order for TPM teams to successfully carry out the requirements of Step 3, the TPM steering committee must make sure the following system infrastructures are in place:

1. An operator and technician maintenance training system.
2. An overdue PM visibility system, and a process for dealing with it.
3. A 5S shopfloor organization system, including a maintenance-tool storage system.
4. Standard times for PMs, and a measurement and reporting system.
5. A "Partner and Practice" system.
6. A visual maintenance documentation system.
7. A 1-Point Lesson system.
8. A machine-performance metric measuring and reporting system.
9. Changes in evaluation, ranking, and reward systems to support wanted behaviors and stop rewarding old, unwanted behaviors. For example, reward those people who engage in disciplined maintenance practices and those who contribute multiple improvements to maintenance plans.

STEP 3 DELIVERABLES

By the time a TPM team has finished Step 3, the maintenance plan that was created to maintain the machine in Step 2 is being carried out with precision by everyone involved in the maintenance plan. People are able to carry out the maintenance plans such that the machine cannot tell the difference between one performer and another. The machine is receiving exactly the maintenance that was planned for it. People are learning to carry out operating and maintenance plans with discipline and precision, and they pay attention to small details.

The machine performance metrics should be improving significantly over their levels at the end of TPM Step 2 activities.

STEP 3'S MINDSET CHANGE
Maintenance must be carried out precisely to be effective and to be effectively improved. Precision maintenance requires a great deal of preparation and support that most maintenance departments lack. If precision is not achieved, further improvements to the equipment maintenance plan will not be effective.

7

Step 4: Prevent Recurring Machine Failures

This chapter describes how to implement
TPM Step 4, including:

- Evaluating PMs for possible improvements
- Creating failure-prevention trouble lists
- Using failure-analysis tools

7

STEP 4 GOALS

In Step 1, activities were performed to restore equipment to "new" condition by making all areas of the equipment clean and free of humanly detectable minor defects.

In Step 2, preventive maintenance (PM) plans were identified, organized, and scheduled.

In Step 3, the goal was to achieve PM implementations consistent with the documented maintenance procedures developed in Step 2. This was achieved when "the machine did not know the difference" as to who was performing the maintenance work.

The goals for TPM Step 4 are to:

- Make PMs faster and easier
- Prevent equipment breakdowns
- Elevate people's knowledge and skills regarding equipment maintenance

In this step, the TPM team will continually improve on the maintenance plans they have been carrying out so diligently in Step 3. Machines are still breaking down, so the PM plan of record and, perhaps, the machine design need some improvement. This step is the heart of TPM activity—the one that the previous steps have been preparing the teams for. Now the teams are going to make all of their PMs routine and reliable, and develop them and the equipment to the point where all machine breakdowns are prevented.

Two principal methods of continuous machine improvement are used in this step—PM evaluations and failure analysis.

PM EVALUATIONS

In Step 3, technicians and operators continually advanced their abilities to carry out their maintenance work with precision at Agilent.

- PMs are conducted with written PM checklists in hand.
- Completed items are checked with a red pen. (If an incomplete PM must be passed on to the next shift of techs to complete, the checklist clearly shows which line items have and have not been completed.)

Checklists help people carry out their PMs with disciplined precision. In Step 4, an element of creative improvement is added to that precision.

Each time a PM is carried out, it should be reviewed for possible improvements: Was it routine to carry out? Were the results reliable? Can it be made easier, faster, or better? Can the checklist items be more precise? Are there missing specs or part numbers? Are procedures clear and unambiguous, so that every tech performs them in the same manner?

The written PM checklist and the red pen also serve a purpose beyond marking completed items. Since technicians have agreed to carry out all PMs precisely, as the checklist describes, they cannot implement changes to the PM at that time. But they can mark up the checklists with their red pens, noting ideas for improvements to the PM. Then, at the completion of the PM, the team can hold a briefing to discuss potential improvements. Once the TPM team reviews the ideas, changes in the PM checklist can be made so everyone doing this PM in the future will perform it with its new improvements. This system retains *disciplined* precision maintenance while providing for *creative* improvement in PM executions. 1-Point Lessons are often written as part of these PM improvements.

Rebuild and Swap Technique

One of the primary improvement strategies to emerge from PM evaluations at Agilent is a "rebuild and swap" technique. A PM, which required a throttle valve on a vacuum machine to be rebuilt, took about four hours with the machine down. Techs were rebuilding the throttle valve on top of a toolbox in the IC fab, wearing clean-room garments and two pairs of clean-room gloves. Anyone who has ever performed maintenance work under harsh conditions can imagine how difficult and miserable this work can be. The PM went well, except one bearing that needed to be replaced could not be located. (The techs had not collected all the parts required by the PM before it began. They believed the part was available in one of their "hidden stocks" in the clean room. It was not.) The old bearing—which was in poor condition—had to be re-installed in the throttle valve. So, even after all of this very difficult work by the maintenance techs, the throttle valve PM was carried out incompletely.

Two weeks later, the throttle valve bearing failed, and the PM had to be repeated. Despite the great effort put into completing this PM, it was unable to prevent throttle valve failure. The technicians decided this PM plan was unsatisfactory and needed to be changed.

The equipment in Agilent's IC fab contained nearly 20 of these throttle valves. They were of modest cost, so the techs obtained an extra valve to use as a "swap" unit. Instead of the PM calling for rebuilding the valve while the machine was down, it was changed to call for the valve to be replaced with the "swap" valve. This procedure was easily accomplished in about 10 minutes. The old valve was then rebuilt on a workbench in the tech maintenance area outside of the clean room, where techs could service the valve in relative comfort. They also had time to acquire any part that was missing because the machine was not down while the valve was being rebuilt. This "rebuild and swap" strategy has been applied to many subassemblies in nearly all of Agilent's equipment. It has greatly reduced machine downtime because parts can be swapped faster than they can be rebuilt or repaired. It has made maintenance work significantly easier and more enjoyable for the maintenance techs.

All of this was accomplished because a TPM team evaluated a PM after completing it and decided to improve it.

Preventive maintenance routines not only must be carried out with precision discipline, they also must be improved continually. Creative improvements are best captured each time that a PM is executed.

FAILURE ANALYSIS

Failure analysis starts where troubleshooting and repair end. When a machine fails, the goal of troubleshooting is to discover what has failed so the machine can be repaired and then returned to service. But discovering what has failed is only the first step of failure analysis. The process then proceeds to determine *why* the failure occurred and to develop countermeasures that will prevent the failure from recurring. Simply replacing failed parts on a machine will do nothing to prevent the recurrence of this failure. The failure-analysis process and its relative position in the machine repair cycle is illustrated in Figure 7-1.

Replacing failed parts on a machine does nothing to prevent the machine from failing again. Machine repair without failure analysis simply maintains the vicious cycle of breakdown maintenance.

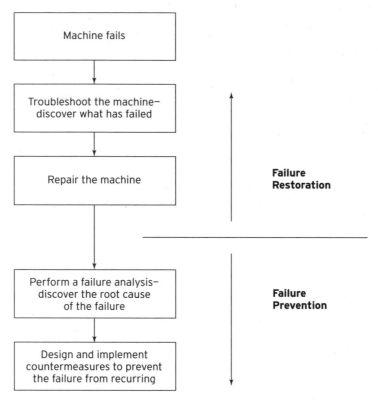

Figure 7-1. The failure analysis process

Failure-Prevention Trouble Lists

A team's first step in preparing for failure-prevention activities is to make a *failure-prevention trouble list*. This is a list of machine failures that must be prevented in the future by TPM team activities. This list will change as problems are resolved and removed from it, and as new failures occur and are added. It will change, too, as people improve their failure-prevention skills and tackle more difficult machine failures. Failure-prevention trouble lists are often simply referred to as trouble lists.

Trouble lists are important because they help make the job of preventing recurring machine failure a scheduled maintenance activity. If failure prevention is only a special TPM project, it will cease when a TPM program is completed. In order to become part of the daily maintenance routine, failure prevention must be supported by the normal factory operating system that schedules all maintenance activities.

Most maintenance work is scheduled by a "work order" concept. Even if actual work orders are not used in a company, the concept of a work order is a good way of thinking about scheduling maintenance work:

- A down machine generates a "work order" for a maintenance tech to repair it.

- A scheduled PM is a "work order" for maintenance techs to perform preventive maintenance work on the machine.
- An open M-tag is yet another "work order" that schedules work for maintenance techs.

However, when the maintenance scheduling system shows no work pending for maintenance techs—machines are running in production, no PMs are scheduled, and no M-tags are open—does this mean that techs can put their feet up on their desks and read the newspaper because there are no open work orders? (Actually, in a breakdown-maintenance culture, this is often the sad truth.)

Every factory's operating system needs a way to schedule the job of preventing recurring machine failures—something that systematically makes this "invisible" work visible. In essence, a trouble list is a work order for TPM action team members to prevent recurring machine failures.

When first preparing trouble lists, it is crucial to decide what kind of machine failures should be included. Most companies either place every machine failure or only the top failure Pareto items on their trouble list. Both of these methods are ineffective in start-up situations. Any organization with a breakdown-maintenance culture will be totally overwhelmed if every machine failure is placed on its trouble list. Likewise, people new to failure-prevention methods probably don't yet have the skills to attack some of the top machine Pareto failures.

Trouble lists must be prepared in accordance with an organization's progress in developing failure-prevention skills. Figure 7-2 illustrates this advancing level of skill.

Improved Failure-Prevention Skills

Start-Up   Practiced   Advanced

Figure 7-2. Advancing levels of failure-prevention skills

Companies with different levels of failure prevention-skills will have different goals for their trouble lists, and the lists will address different types of machine failures.

- *Start-up level.* The goal at this level is to develop failure-prevention skills and advance the maintenance culture away from breakdown maintenance and toward failure prevention as a normal daily activity. At this level, include on the trouble list only those machine failures that have known repairs which effectively restore the failed machine to normal operating service. This is the easiest place to start. It is more difficult to prevent a failure that does not have a known repair. Furthermore, sort the trouble list from easy to hard so that team members can develop their failure-prevention skills on easy problems first.

Often, machine failures can be added to trouble lists at the start-up level only in hindsight. Sometimes a recurring failure gets "fixed" repeatedly, each time in a different way. At the time of the repair, a technician takes the best action known at that time to repair the machine. Later events, however, may indicate the repair did not address the root cause of the failure—the machine continues to break down in the same or a similar manner. In hindsight, once the real root cause is repaired, all these failures can be seen as a single problem for the trouble list rather than a long list of problems.

- *Practiced level.* The goal at this level is to extend failure-prevention activities to machine failures that have not responded well to conventional troubleshooting and repair techniques—those recurring machine failures that have no known effective repair. This will stretch the team's proficiency level in preventing machine failures. Again, sort these problems from easy to hard so new skills can be developed on the easiest problems first

- *Advanced level.* The goal at this level is to prevent all machine failures. Every machine failure should be placed on the trouble list and sorted by business priority.

Failure Prevention Skill Level	Types of Machine Failures to be Placed on Trouble Lists
Start-up	All machine failures—whether unique or recurring—with known repairs
Practiced	Recurring failures that have no known repair
Advanced	All machine failures

Trouble lists can be assembled during regular TPM action team meetings. The recent history of sporadic machine failures should be reviewed and appropriate items added.

TPM teams should keep in mind that problems on their trouble lists at any level might require more skill to prevent than they possess. Whenever this occurs, a special *focus team* should be created to deal with that single machine failure. Focus teams often are engineering centered and might even include members from outside the organization, such as vendor representatives or consultants. A focus team should target the single problem it was chartered to prevent and then disband.

Following are a few examples of machine failures that should be placed on trouble lists at different levels of failure-prevention skill.

Start-up Level: Machine Failures with Known Repairs

A variety of equipment problems in one of Agilent's machines was traced to a worn ribbon cable in the machine's wafer-handling system. The worn ribbon cable was only discovered as the root cause of these failures after many false attempts were undertaken to repair the problems. This particular cable was flexed and rotated extensively during the machine's normal operation, and therefore had a limited life expectancy. Once the machine problems were traced to the failure of the cable, future cable failures needed to be prevented by some sort of countermeasure. (The cable's replacement should not be scheduled by the onset of machine problems caused by its own failure.) In this case, a use-based PM based on the number of wafers handled was set up to replace the cable as it approached the end of its expected life and before any machine performance problems were caused by the deteriorating cable.

Practiced Level: Recurring Machine Failures with No Known Repair

One machine in Agilent's IC fab began to exhibit a new minor stoppage problem by reporting an error code to the operator. The machine operator responded to the stoppage by acknowledging the error and then commanding the machine to continue operation. Several different error codes were displayed at different times, but all had to do with wafer robot problems at the machine's "interface station." Repeated attempts by maintenance technicians to troubleshoot and repair this new sporadic failure could not resolve the problem. After a month of continued machine errors, the problem was added to the trouble list for more rigorous failure-analysis efforts. The failure analysis revealed numerous minor defects in the interface station mechanical, vacuum, and electrical systems. These minor defects were restored to normal, and maintenance plans were put in place to keep the systems in this newly restored state. The machine errors were thus eliminated.

Advanced Level: All Machine Failures

One machine in Agilent's IC fab had a circuit board failure in a power supply. In the five-year history of this type of machine, this board had never failed. Since this was a very rare failure and difficult to diagnose, it would not likely have been placed on a trouble list during earlier levels of failure-prevention efforts. At the advanced level, *all* machine failures are added to the trouble list, and efforts are made to prevent the failure from ever occurring again—the exception, of course, being failures that are designed to be handled with breakdown maintenance. In the case of this failing circuit board, the root cause was that the board was unplugged while it was powered on. Although this event only occurred rarely, a change was made to documented maintenance procedures so it would never be repeated.

Failure-Analysis Tools

Whenever a machine fails, some part of its overall maintenance and operating plan has failed. TPM action teams must analyze the failure and take steps to prevent its recurring. Several failure-analysis tools are available:

- *5-why analysis*. A classical approach to finding the root cause of a failure.
- *Maintenance analysis*. A simple approach to identifying improved maintenance plans. Two specific "whys" are asked. (Note, this is not a short form of 5-why analysis.)
- *P-M analysis*. An advanced approach to resolving infrequent but very difficult equipment failures.
- *P-M Lite*. A simplified approach to P-M analysis—developed at Agilent—that can be used by most technicians to create countermeasures for almost all types of equipment failure.
- *Physical analysis*. A useful, independent failure-analysis tool or a method that can be used in conjunction with any of the above types of failure-analysis methods to improve their effectiveness.
- *PDCA*. A generic problem-solving tool used to resolve any type of problem, whether a machine failure or some other kind.

5-Why Analysis

A 5-why analysis begins with collecting data about the failure from operators and techs. These data include:

- What fails
- Where the failure occurs
- When it occurs—both its current and its historical failures
- How it fails

Armed with these data, a TPM action team can continue its 5-why analysis, (often called a why-why analysis). The procedure is simple. Given the above data, ask why the primary failure symptom occurred. Then ask why that occurred. Ask why that occurred. Repeat this "why" questioning five times or more, until the root cause of the failure is known. Then a countermeasure can be implemented to address the root cause and prevent recurrence of this failure in the future.

For example, one of Agilent's production machines used to drop a wafer about once every three days. We began by recording answers to our most basic questions about this failure over a period of time. The results of this data collection were as follows.

- *What fails?* The vacuum arm is unable to hold on to the wafer.
- *Where does this failure occur?* When the wafer handler is rotated into a position in front of chamber #2.

- *When does this occur?* Every two or three days, with no obvious schedule or pattern.
- *How does this occur?* The wafer simply falls off the blade of the wafer-handling arm as if there were insufficient vacuum at the blade—which is what holds on to the wafer. There is nothing touching the wafer in the location where it gets repeatedly dropped.

Then the "why" questioning begins:

1. *Why did the machine stop running?* Because the wafer fell off the wafer-handling blade.
2. *Why did the wafer fall off of the handling blade?* Because there was insufficient vacuum at the tip of the blade to hold onto the wafer.
3. *Why is there insufficient vacuum at the tip of the blade?* Direct observations of this machine operation—even when it was not dropping wafers—revealed that, in this particular position, the coiled vacuum hose to the blade was being pinched.
4. *Why does this hose get pinched in this particular position?* Because the hose is rubbing against some wires that are not securely tied into place.
5. *Why are these wires unsecured?* The wire tie wraps are cut every three months when a PM is performed, because the wires complicate the access required for the technician to replace a part. Because the technician knew that this PM would be continually repeated, he did not bother to retie the wires into place with tie wraps.

Once five "whys" have been asked and answered, the root cause of the problem is usually revealed. And once we understand the root cause of the problem, the countermeasure to prevent the problem from recurring is often obvious.

In this case, we resolved the problem by tie-wrapping the wires in a secure manner, out of the way of the coiled vacuum hose so that the wire would never catch the hose and cause it to be pinched. A line item was added to the PM checklist requiring the wires to be resecured in their correct position with tie wraps. A 1-Point Lesson was also prepared to teach technicians who perform service on this machine about the importance of tying each of the wires securely in its proper location.

A side effect of this investigation was that technicians also noticed that the coiled vacuum hose was somewhat deteriorated, and there was no specified PM to inspect or service this hose. These hoses were simply being replaced when they failed catastrophically. So a hose inspection and conditional replacement PM was also established. These two solutions completely eliminated the wafer dropping problem.

In another example, wafers were being damaged in a machine that processed them in a vacuum chamber. The damage was so slight that it generally remained unnoticed until the wafers reached the wafer test station at the end of the production process. By this time, many wafers were damaged, at great expense. The following was our 5-why analysis for this situation.

1. *Why are the wafers being damaged?* They are being scratched by the wafer-handling robot.
2. *Why is the robot scratching the wafers?* The robot has slipped out of its critically aligned position.
3. *Why is the robot misaligned?* A bolt which holds one of the robot's critical alignment adjustments is vibrating loose.
4. *What causes the bolt to become loose?* The bolted joint is designed so that the bolt cannot pull a tight metal-to-metal joint. Stepper motors, which drive the robot, induce a vibration in the bolted joint when the robot is in motion. Also, because of the susceptibility of the machine's wafer process, thread-locking chemicals cannot be used to lock the bolt threads against vibration.
5. *Despite these challenges, why can't we keep this bolt from working loose?* Figure 7-3 shows the bolted joint, which takes a very long time to work loose. Our experience is that it will remain tight for over a year. Our solution to prevent the bolt from vibrating loose was to retorque the bolt to its proper tightness in a timely manner—about once every six months.

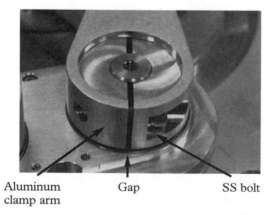

Aluminum clamp arm   Gap   SS bolt

Figure 7-3. A joint that repeatedly failed by bolt loosening

This area of the machine was accessed about once a month, and a semi-annual PM already existed, so adding a checklist item to re-torque the bolt was quite simple. It also took less than a minute to perform, so this countermeasure to the problem required trivial effort.

The real challenge to implementing this countermeasure was to determine the proper torque value for the bolt. Torque charts for ordinary bolt joints offered conflicting advice. Should we use the torque value for a stainless steel fastener or the torque value for an aluminum thread? Would either of these torque values be correct, considering the "flexibility" of the joint due to the gap?

Rather than guessing at this important value, we learned the correct torque value by creating a test situation with the same materials and similar geometry as the bolted joint.

Figures 7-4 and 7-5 show the test parts we created.

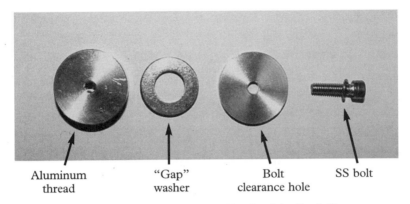

Aluminum thread   "Gap" washer   Bolt clearance hole   SS bolt

Figure 7-4. Test pieces were fabricated to simulate the failing bolted joint

"Gap" simulator

Aluminum thread

Figure 7-5. Test pieces assembled for bolt torque tests

The test pieces were assembled to create a situation for the fastener with requirements similar to those in the actual robot joint. The washer created a simulated gap between the two metal pieces that the bolt was pulling together.

The test assembly was placed in a vise so that the bolt torque tests could be performed. The bolt was torqued to a low value, then disassembled; the bolt and aluminum threads were inspected. The test was repeated at higher torque values until the assembly failed on multiple bolts. The failure in each case was the bolt itself yielding and then shearing into two pieces. The aluminum thread remained undamaged. The bolt began to yield at about 45 inch-pounds of torque. A tightening torque for this bolt of 70 percent of the failure value was established—about 31 to 32 inch-pounds. (See Figures 7-6 and 7-7.)

Figure 7-6. Bolt torque being measured at joint failure

Figure 7-7. Failed test bolts

This failure analysis demonstrates that finding the solution to a problem often requires learning something in detail about the failing machine components. In this case, we needed to understand the correct tightening torque value of the bolt and the frequency with which it needed to be retightened.

Guessing Wastes Resources

5-why analysis is, perhaps, the most common failure-analysis method used today by technicians who are trying to prevent recurring machine failures. Agilent has found it a useful tool but not a perfect one—a 5-why analysis sometimes fails when a technician cannot provide an answer to one of the "whys." For example, let's suppose that a bearing failed and caused a machine shaft to freeze:

1. *Why did the machine stop running?* Because a bearing froze.
2. *Why did the bearing freeze?* I don't know.

The person or team performing this failure analysis is now stuck. They are not bearing experts, don't have a bearing failure-analysis laboratory available, and any answer to this second "why" is just a guess. Once we start guessing at answers, any countermeasures that we design are based on mere speculation—we do not know the true root cause of the problem.

In this case, we might assume that the bearing failed for lack of proper lubrication, even though we had already lubricated it based on the machine manufacturer's recommended schedule. We might then inappropriately overlubricate the bearing in an ignorant attempt to prevent it from failing, when the root cause might have been vibration due to an unbalanced rotating load. Overlubrication will not help prevent this bearing failure at all, and may, in fact, make it worse.

Sometimes "educated" guesses in a 5-why analysis do produce a reasonable failure countermeasure, but most often, a team wastes resources implementing countermeasures that do not work.

In a case where 5-why analysis doesn't help the team find the true root cause of a problem, it is best to use other failure-analysis methods.

Maintenance Analysis

Maintenance analysis is not a short form of 5-why analysis, even though it might appear so at first glance. Rather, it is a simple approach to designing a maintenance solution to a machine failure. Two simple "whys" are asked and answered:

1. *Why did we not see this failure coming?* Many failures provide early warning signs, some of which may appear weeks or months ahead of the failure. Others may appear only hours or days ahead. No matter the time frame, these warning signs are often detectable by human senses—perhaps as much as 90 percent of the time, and three-quarters of these are detectable visually. Other failures occur at predictable intervals of part age or use. What PMs might be put into place to catch the early warning signs of this failure?[1]
2. *Why did our maintenance plan not prevent this failure from occurring altogether?*
   - Did the part die earlier than its expected life because of accelerated deterioration? If so, why are we not maintaining the proper conditions-of-use for this part?[2]
   - Did the part reach its natural life expectancy? If the failure was a true end-of-life failure, and the life expectancy is predictable, a PM could be created

---

[1]For more information, see "Graceful Deterioration" on page 120.
[2]For more information, see "Condition-of-Use and Life Analysis" on page 222.

to replace the part when it is approaching the end of its useful life, but before it fails.

The following is an example of a maintenance analysis performed by Agilent technicians on one of our IC fab machines.

A metal bellows was used to provide a high-vacuum seal around a shaft that slid through the wall of a vacuum chamber (Figure 7-8). The bellows failed when a crack developed in the wall, and atmospheric air entered the vacuum chamber, which destroyed the chamber vacuum and all the wafers that were being processed in it at the time.

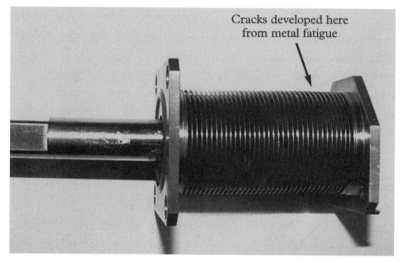

Figure 7-8. A vacuum bellows assembly

Why did we not see this failure coming? The bellows seal deteriorated by fatigue in the metal as it was flexed during the operation of the machine. However, this metal fatigue was not measurable by any means practical for our maintenance department. Therefore, inspecting the bellows did not provide any insight into any pending failure of the bellows.

Why did our current PMs not prevent this failure?

- Did we fail to provide the necessary conditions-of-use for the bellows, causing it to die early? No. The bellows operated a natural lifetime in the environment for which it was designed, so no accelerated deterioration was taking place.
- Was the natural life of the bellows predictable? Yes. The manufacturer of the bellows understood its fatigue failure very well and could predict the number of cycles the bellows was likely to last in any given design situation. By inquiring, we found that the anticipated life of the bellows in our

situation was about 500,000 cycles. Since the bellows cycled twice for every wafer processed, the life expectancy of the bellows was about 250,000 wafers processed through the machine. This took several years to occur, and because the failure was so infrequent, we had never created a bellows replacement PM to prevent the bellows from failing.

Our maintenance solution to prevent this failure from recurring was to create a PM to replace the bellows every time 225,000 wafers were processed through the machine. This moderately priced bellows could be replaced in about 30 minutes. This was far less expensive than an unpredictable bellows failure, which often scrapped all the production wafers that were in the vacuum chamber when the failure occurred.

Maintenance analysis is a very good failure-analysis method. Still, it has obvious limitations—the most notable being that only maintenance countermeasures are being sought. In many cases of machine failure, other countermeasures are required to prevent recurring failure, like increased knowledge or skill for operators or techs, or a design improvement in some weak area of the machine. Maintenance analysis does not always provide for a true root-cause understanding of the problem, nor for a wide range of countermeasures to be implemented to resolve the problem.

P-M Analysis

P-M analysis is a technique that has already been documented in other TPM textbooks; it can be an extremely powerful and effective tool for preventing recurring machine failures. P-M does not stand for planned maintenance, as might be supposed; the P stands for physical and phenomena, while the M stands for the four sources of machine conditions: man, method, material, and machine.

Unfortunately, P-M analysis has a couple of drawbacks that limit its widespread distribution as a failure-analysis tool. First, P-M analysis is only recommended for problems that occur less than .5 percent of the time—that is, very intermittently. For failures that occur more frequently, more conventional approaches to failure analysis, like 5-why or maintenance analysis, will produce faster improvements in machine performance with less effort.

Second, P-M analysis is difficult for many machine technicians to master, and many engineers find it difficult to implement. The problem is the very first step, which requires a description of the physical phenomena of the proper operation of the machine part. There is only one technically correct description, regardless of how it is worded.

For example, suppose I am analyzing the continual failure of a light bulb filament. I might describe the physical phenomena of a light bulb as "visible light resulting from heat generation that occurs in a resistive filament wire as the result of an electrical current passing through it, this light being produced once the filament temperature rises above 1,000 degrees centigrade."

This is a not the way that most maintenance technicians perceive a light bulb. If the light bulb in their shop lamp is burning out too often, they will probably replace it with a long-lasting rough-service bulb; most of them will never ponder the physical phenomena of the light bulb itself. Very few techs—or engineers—investigating the root causes of a machine failure can provide a technical description of the physical phenomena of the part. However, the description of the physical phenomena is the first step of P-M analysis and lays the foundation for all the remaining steps. If the first step is poorly stated, the remaining steps will be built on a poor foundation, which may produce insufficient countermeasures to resolve the failure.

P-M analysis should not be rejected just because of its complexity; it has been a very useful and powerful tool at Agilent in a few difficult failure situations. However, it is not our primary failure-analysis tool—we use it when other methods fail us.[3]

P-M Lite

P-M Lite is the name coined by Agilent for a failure-analysis tool developed in our factory that is based on the principles of P-M analysis and other TPM methods. Most maintenance technicians and engineers learn this relatively simple method easily, and it has resolved about 90 percent of the machine failures to which it has been applied.

In the few cases where it does not provide a complete solution for preventing 100 percent of a particular machine failure, we augment it with physical analysis (described later in this chapter), or we use P-M analysis. But P-M Lite is so easy to use and so successful at preventing machine failures that it is our most commonly used failure-analysis method.

P-M Lite consists of five steps:

1. Identify all the factors at the machine—hardware and other conditions—that can possibly influence the failure.
2. Define the optimal condition for each of these factors.
3. Compare the actual conditions of these factors to their optimal conditions. Restore all abnormalities found.
4. Create maintenance and training plans to keep these conditions at their optimal point. This will assure natural lifetimes of all the machine components involved. (Note: This step is actually a condition-of-use analysis.[4])
5. Create either time-based, use-based, or condition-based maintenance plans to replace or service parts as they approach the end of their natural lives.

---

[3]For more information on P-M analysis, see *P-M Analysis* by Kunio Shirose, published by Productivity, Inc. (ISBN 1-56327-035-8).

[4]For more information on the concept of maintaining component conditions with this technique, refer to "Condition-of-Use Analysis" page 222.

To more fully explain these steps, an example of this type of failure analysis follows.

Wafer-handling failures were occurring in a very complex Agilent production machine that had been in service for several years. None of these failures occurred when the machine was new. Repeated investigations into the failures by technicians and engineers failed to find the root cause of the wafer-handling difficulties. No trouble could be found in the machine. However, every couple of days, the wafer handler would mishandle a wafer, and the machine would stop operating, requiring a technician to restart it. The following failure analysis was performed.

Step 1. Identify all the factors at the machine—hardware and other conditions—that can possibly affect the failure.

- Technicians who were familiar with this machine examined it carefully and made a list of all the machine subassemblies that had any effect whatsoever on wafer handling.
- They identified the contributing system factors as:
  - The slit valves
  - The wafer-handling robot
  - The I/O valve
  - The storage elevator
  - The cassette lift assembly
  - The chamber lift assemblies
  - The pneumatic control valves

Step 2. Define the optimal condition for each of these factors.

- Figure 7-9 shows a slit valve actuator and lists the optimal conditions that should exist in every slit valve assembly at all times to make sure this subassembly is contributing properly to the handling of wafers in the machine.

Step 3. Compare the actual conditions of these factors to their optimal conditions. Restore all abnormalities found.

- Figure 7-9 also lists the abnormal conditions actually found in various slit valve assemblies. No fewer than four minor abnormalities were found in any one slit valve, but none of them seemed significant. The technicians tending this machine found that the slit valves were all working properly and passed functional testing—they opened and closed properly on command. Still, all the abnormalities found were restored to their optimal state before being returned to the machine.

Step 4. Create maintenance and training plans to keep these conditions at their optimal point.

- Several 1-Point Lessons were created to teach equipment technicians how to inspect all the slit valve conditions inside of the machine for their optimal state. Semiannual inspection and service PMs were created to ensure that all the optimal conditions were maintained, including optimal opening and closing speeds to prevent valve "slamming." These maintenance procedures were designed to ensure that the slit valves were properly maintained during their natural lifetimes.

Step 5. Create a restoration maintenance schedule to replace parts as they approach the end of their lives.

- Based on past performance of the equipment, wafer-handling problems began after the machine passed two years of age. With no other data available, a biannual PM to rebuild each slit valve was created. This PM included replacing the pneumatic cylinder and all the vacuum seals. A rebuild kit was created, containing all the parts required. A visual rebuild procedure was also created to make certain that the maintenance work would be performed precisely as described, no matter who did it in the future.

The left column in Figure 7-9 lists the optimal conditions that should exist in every slit valve assembly at all times to make sure this subassembly is contributing properly to the handling of wafers in the machine. The right column lists the abnormal conditions actually found in various slit valve assemblies.

Steps 2 through 5 were then repeated for each of the other wafer-handling subassemblies identified in Step 1 of the P-M Lite procedure. In this way, all minor abnormalities in all subassemblies were corrected. None of these steps proved to be challenging for the technicians working on this failure analysis, as the work utilized maintenance skills they have mastered. The work simply took some time to complete. As can be seen in Figure 7-10, the wafer-handling failure rate was reduced to zero with this P-M Lite failure-analysis procedure.

Interestingly, each of the abnormalities found in all of the machine subassemblies appeared insignificant. Every subassembly appeared to work fine even before the abnormalities were removed. None of the technicians or engineers could link any abnormality they found directly to any specific machine problem. However, the sum of these minor abnormalities obviously created the machine's wafer-handling problems because eliminating the minor abnormalities eliminated all of the wafer-handling failures. This is a clear demonstration of the effects that even seemingly insignificant minor defects can have on the performance of a machine.

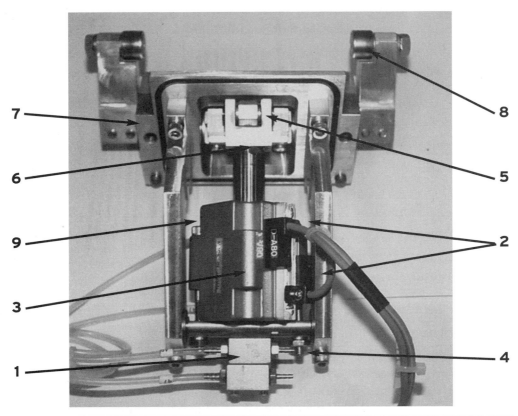

Item	Optimal Condition Required	Actual Condition Found
1	Speed controls installed correctly to regulate pneumatic cylinder exhaust	Speed control valves installed backward
2	Each sensor located at a specified location	Piston location sensors installed in wrong position
3	Piston seals must be leak-free and tight	Cylinder piston seals leaky and loose
4	Bolts must be properly installed, torqued, and locked	Loose or missing bolts
5	One specific shim must be placed on each side of the yoke	Center shims missing—cylinder not perpendicular to yoke
6	Yoke must be installed on cylinder rod thread with specified threadlocker	Yoke unthreading from cylinder and striking machine surfaces
7	Vacuum seals must be leak-free	Vacuum seals leaking
8	Cam followers must be adjusted to a specific clearance	Cam followers misadjusted
9	Sensors should be on left side	Cylinder 90° out of rotation

Figure 7-9. Steps 2 and 3 of P-M Lite demonstrated on a slit valve assembly

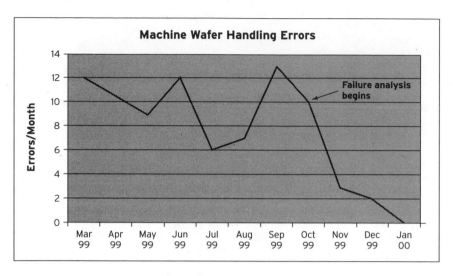

Figure 7-10. The results of P-M Lite failure analysis on wafer-handling failures in an Agilent Production machine

Physical Analysis

Physical analysis involves a detailed physical inspection of machine components and a detailed investigation of how a machine subassembly normally works. It may also include a detailed "autopsy" of failed hardware in an effort to determine the root cause of the failure. This method is often used in conjunction with other failure-analysis tools, but it also can be useful as an independent failure-analysis method.

Physical analysis consists of two investigative steps followed by standard root-cause analysis, countermeasure design, and implementation of the solution. These investigative steps often lead to understanding the root cause of the failure:

1. *Conduct a physical analysis of the machine.*
   a) What part or adjustment failed?
   b) How does the failure affect the machine operation?
   c) Describe the repair used to restore the machine to operating service. Were parts replaced? Were adjustments made?
   d) Conduct a physical analysis of the operating hardware associated with the failure. Describe in detail how the entire system containing the failure normally works. Be thorough—understand the entire subsystem completely.

2. *Conduct a physical analysis of the actual failed parts, if available.*
   a) Can you determine *specifically* where the failure occurred in the part?
   b) Can you determine *specifically* how the failure occurred in the part?
   c) Can you determine *specifically* why the failure occurred in the part?

In many cases, successful completion of one or both of these two steps provides enough insight into the root cause of the failure to design appropriate countermeasures. Agilent's experience with this technique is, surprisingly, that learning in detail how the entire assembly containing the failure is supposed to normally work (Step 1 of physical analysis) can often reveal more about the failure than examining the failed part (Step 2 of physical analysis). In other words, before a team can understand how a machine or machine component fails, it is necessary to first understand, in detail, exactly how it is supposed to work.

Agilent equipment technicians sometimes tended to overlook this important step. When we started focusing our understanding on how the system was supposed to normally work, we discovered this was often the breakthrough needed to understand the root cause of the failure and design preventive countermeasures. An example follows.

Agilent's IC fab contains over 200 vacuum pumps that support the operation of many IC production machines. A dedicated crew of vacuum-pump maintenance technicians carries out rigorous preventive maintenance on all of these pumps. Yet, one particular type of pump was failing very frequently, and when it did, technicians replaced it with a rebuilt pump and sent the old one to a vacuum-pump rebuilder. The company rebuilding the failed pumps could only tell us that the pumps failed from overheating. The pumps had a thermostat that limited cooling water flow to control pump temperature. This thermostat had always been replaced as part of the rebuilding service provided by the pump rebuilder. Still, the pumps continued to fail at a high rate.

Agilent pump technicians decided to perform a physical analysis on a pump cooling control valve. They performed this analysis on a *new* valve—not a failed one, so they used only Step 1 of physical analysis. They were trying to understand specifically how the valve normally worked. They quickly understood the basic principles of the valve, but found it much more difficult to describe *specifically* how the cooling water thermostat worked in minute detail. It contained many different parts, chambers, and cooling lines going to various parts of the pump. No detailed schematic of this pump subsystem was included in the pump maintenance literature.

The pump-mounted valve is shown in Figure 7-11. The technicians became more familiar with the valve by disassembling it and tracing the cooling water path through every portion of the valve. This enabled them to draw a complete schematic of the valve's cooling water circuits. They were able to identify every part on the schematic with the actual part or chamber in the valve body, and they knew exactly how the hardware in the cooling circuit operated.

Once the techs understood how the cooling circuit worked, they could more easily understand how it might fail. They already knew that the pumps were failing by overheating, and this meant that too little cooling water was flowing to at least one of the cooling circuits controlled by the cooling valve. They found that some of these cooling lines were fed through reduced-size ports inside of the valve body, and behind one of these ports was a fine mesh filter screen buried deep inside the valve (Figures 7-12 and 7-13). Any foreign contamination in the cooling water could easily collect in this screen and eventually reduce the flow to parts of the vacuum pump.

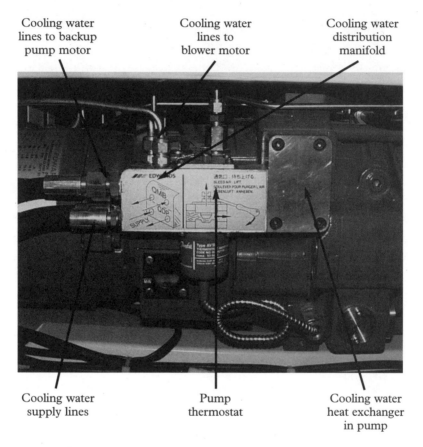

Cooling water lines to backup pump motor

Cooling water lines to blower motor

Cooling water distribution manifold

Cooling water supply lines

Pump thermostat

Cooling water heat exchanger in pump

Figure 7-11. Vacuum pump distribution manifold and cooling water thermostat

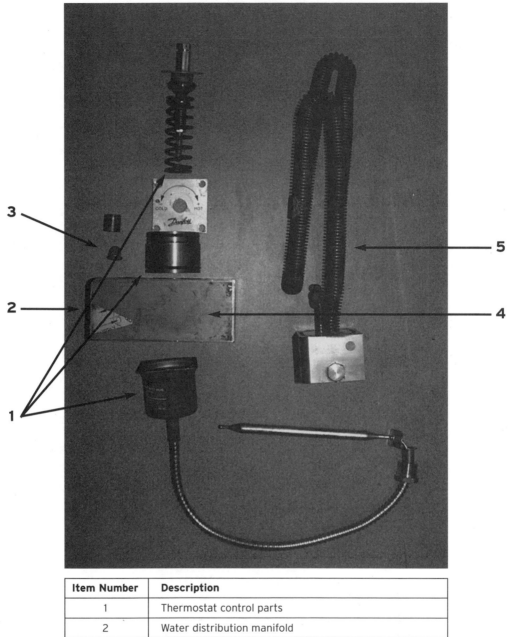

Item Number	Description
1	Thermostat control parts
2	Water distribution manifold
3	Water filter screen
4	Thermostat body
5	Vacuum pump cooling heat exchanger

Figure 7-12. Disassembled cooling water control valve

Filter screen
hidden inside
of valve

Figure 7-13. Cooling manifold ports and location of internal filter screen

Their physical analysis of hardware in the cooling circuit allowed Agilent technicians to understand in detail how the pump cooling circuit operated, which gave them insight into how it might cause pump overheating. They suspected that the small filter screen inside of the water distribution manifold might be plugging up with contaminants in the cooling water and reducing the water flow.

They did not have a failed pump to inspect, so they tested their hypothesis in another way. They placed a filter in the cooling water supply line feeding one pump to trap any contamination from the water. They carefully inspected the filter every two weeks and examined the material trapped by the filter. They found small contaminants in the cooling water that were large enough to get trapped by the screen inside the valve. They believed these particles were the source of their pump overheating failures because their filter collected a sizeable amount of them in a surprisingly short amount of time. The pump rebuilder never discovered the plugged-up screens because the entire cooling water control valve was replaced when it failed—no one disassembled it. None of our pump technicians even knew the screen was present in the valve body until they performed a physical analysis on the valve.

It was impractical to place a filter on the cooling water line of each of several hundred pumps to resolve the problem. Instead, technicians added improved filtration, water treatment, and maintenance plans to the entire cooling water system. Their failure hypothesis proved to be correct. No pump failed from overheating after the new maintenance plans removed contaminants from the cooling water.

The successful countermeasure to this very expensive pump failure was made possible primarily by understanding how the pump cooling circuit was supposed to work. Before this physical analysis was conducted, the pump maintenance techs had only a vague notion that the cooling valve included a thermostat that somehow controlled the temperature of the cooling water to various portions of the pump.

 Vague notions about a machine's normal operation are not sufficient to help technicians prevent machine failures.

A detailed working knowledge of the machine hardware in the failing subassembly is required and can be obtained by careful physical analysis of both new and failed hardware. Books and drawings supplied by the machine manufacturer are helpful, but a personal physical inspection of the hardware is absolutely necessary to perform a successful failure analysis of this type.

PDCA

PDCA—the acronym for plan, do, check, and act—is a general industry term used to describe a basic improvement cycle. The cycle is most often described in seven general steps:

Plan
1. Select the issue to be improved.
2. Analyze the current situation and set an improvement target.
3. Analyze the root cause and determine the corrective action that needs to be taken.

Do
4. Implement a prototype solution.

Check
4. Evaluate the results.

Act
5. Take appropriate action—implement, standardize, and document the new process. Train employees.
6. Make future plans—continue with this issue or select a new issue.

The following instruction set describes these steps more fully.

1. *Select the issue to be improved.*
    - What indicator changed?
    - What needs improving?
    - State why the issue is important.
    - Identify the factory measure this issue affects:
      – Cycle time

- Throughput
- Defect density
- Yield
- Cost
- Productivity, etc.

- Once the problem is selected, the owner should decide what resources are needed to tackle the next few steps in the PDCA cycle. If a team is needed, identify its members by name.

(Note: anyone can initiate a PDCA to capture a problem, but the problem discoverer does not have to become the problem owner.)

2. *Analyze the current situation.*
   - What?
     - Describe the problem in detail. Include evidence germane to the problem.
     - Describe the situation that needs improving. Identify the critical measure(s).
   - Where?
     - Identify the area in the factory, the equipment, and/or the part of the organization involved.
   - When?
     - Include a history of the situation.
   - Who?
     - Include shift information, people who know about this situation, and who discovered it.
     - Set an improvement goal for the measure.

3. *Analyze the root cause of the problem.*
   - Use a 5-why analysis or other failure-analysis method to try to understand the root cause of the problem. Other tools may also be helpful:
     - Histograms
     - Pareto charts
     - Trend charts
     - Cause-and-effect diagrams (fishbones)
     - Check sheets
     - Flow charts
     - Control charts
     - Scatter diagrams or any other form of data plot
     - Experiments

4. *Implement a prototype solution.*
   - Solutions to the problem will relate to the root-cause analysis. The purpose of a solution is to either control or detect the problem:
     - "Control" solutions are proactive and designed to prevent the problem from occurring

- "Detect" solutions cannot prevent the problem but stop the manufacturing process when the problem is detected
- For a documentation problem, a solution requires two things:
  - Officially released document
  - Proper training
  - Solutions may be:
  - Administrative solutions—new procedures to prevent a future occurrence of the problem
  - Engineered solutions—new equipment designs which prevent a future occurrence of the problem
  - Combinations of engineered and administrative solutions
- Evaluate alternative solutions for:
  - Effectiveness/robustness
  - Cost
  - Timeliness
  - The most appropriate solution from the team's perspective
- Design a test of the proposed solution. Outline the test procedures:
  - What tasks need to be performed?
  - Who owns them?
  - What is the schedule for the test?
  - Determine the success criteria for the test—how will the team know if the solution is effective?
- Does this plan correct the root cause?
- Identify the customer—the person representing the area affected by this issue—and schedule a checkpoint to review the solution plan.
- Assess team needs to determine if additional resources are necessary to proceed successfully.

5. *Evaluate the results.*
   - Test the proposed solution.
   - Compare before-and-after data.
   - Was the test successful?
   - Were the success criteria satisfied by the solution?
   - Do results show the root cause has been eliminated?
   - Does this solution guarantee that the problem will not reoccur?
   - Are resources available to implement this solution?
   - Does the customer agree with the plan to implement this solution?

6. *Take appropriate action to implement changes.*
   - Change hardware if an engineering change is being made
   - Use the change process for affected documentation
   - Plan personnel training where required
   - Do issues remain? What plans are in place to address them?

7. *Make future plans.*
- Apply the solution to all identical equipment.
- Apply this solution to other factory areas with similar problems.

   Although these general instructions make PDCAs seem quite complex, the vast majority of them really are not. Figure 7-14 shows a relatively simple, but typical, PDCA.

PDCA PDCA Log # Etch-2706	**(1a) Issue Statement/Why Selected** Improve operators' ability to inspect components inside the irradiator boxes on the bake machines. [This is important because weekly inspections are very difficult and time consuming to complete.]
	(1b) Champion/Team Members/Date Initiated Jim Leflar/Todd Cito–September 18, 1999

(2) Current Situation • The C&I standard calls for weekly operator inspections inside the irradiator box. • The irradiator box cover is extremely cumbersome to remove. • Four bolts hold the cover in place, and tools are required to remove them. • The sheetmetal cover is extremely close fitting and is very difficult to remove and re-install, even after the bolts are removed.	**(5) Results** • A new cover was designed and reviewed by the machine operators. • A prototype was constructed and installed on one machine. The initial installation was somewhat difficult because of the design location of mounting holes. • Operators found the design highly effective but suggested the removal of an extra inspection door that was added to the design. • Design changes were made according to this evaluation and the prototype modified. • The final design was found to be acceptable.
(3) Root-Cause Analysis • The cover design makes the cover inherently difficult to remove, considering our maintenance needs for weekly operator inspections behind the cover.	**(6) Document/Implement** • Cover panels for the remaining six bake machines were fabricated and installed. • A 1-Point Lesson was created and distributed to teach operators how to access the machine through the new cover.
(4a) Solution/Testing Plan • Redesign the cover and build a prototype. • The new cover will require no tools for removal or installation. • The new cover will easily be removed and replaced by an operator in an ergonomically sound fashion in no more than a few seconds.	**(7) Standardize/Make Future Plans** • Cover panels on any machine that are difficult for operators or techs to remove for inspection or maintenance purposes are being scrutinized to see how speed of access can be improved by any means–including redesign of the covers.
(4b) Scheduled Checkpoints **Date** #1 Customer agreement to proposal __9/27/99__ #2 Identify necessary resources __9/28/99__ #3 Customer agreement to final solution __11/08/99_	

Figure 7-14. A simple PDCA

Countermeasure Plans

No matter what failure-analysis method is implemented, a broad range of countermeasures that can prevent recurring machine failure should always be considered. All these countermeasures fall into two broad categories:

- Discover and create the optimal equipment state.
- Advance people's knowledge and skill so they can maintain the equipment in this ideal state.

Types of countermeasures to consider when trying to prevent a machine failure from recurring include:

- Changes in technician-scheduled maintenance plans
- Changes in the operator cleaning and inspection standard
- New or improved visual machine or process controls
- Mistake-proof designs or procedures
- 1-Point Lessons or changes in documented procedures for either operators or techs that teach them something new and relevant about the machine—either what they should be doing or, perhaps, what they should not be doing
- Additional training materials or classes
- Changes in maintenance tools
- Changes in operating procedures
- Changes in manufacturing process design
- Changes in weak machine designs
- Acquisition of new maintenance technology that allows technicians to detect machine conditions beyond what they can discern with their human senses.[5]

---

[5]For a more thorough description, see "Extended Condition Monitoring" on page 234.

STEP 4 MASTER CHECKLIST

As a TPM team advances through the activities of Step 4, they should refer to the following checklist to be certain they are progressing successfully.

STEP 4 MASTER CHECKLIST: PREVENT RECURRING MACHINE FAILURES
A PM evaluation is part of completing every PM. The goal is making every PM easier, faster, and better
☐ Upon completion of each PM, the tech redlines proposed improvements to the PM checklist and procedures
☐ The TPM team reviews the proposed changes
☐ The maintenance documents are changed through the maintenance change process, and a change notice is sent to all affected operators and technicians
☐ All affected people receive appropriate training in new maintenance procedures
A failure analysis is completed for each machine failure:
☐ A trouble list contains up-to-date machine failures that must be prevented from recurring
☐ A status report lists all failure analyses pending, in process, and completed
☐ The countermeasures implemented in each case are recorded
☐ The results of the countermeasures in each case are recorded
The knowledge of "hundreds of minor equipment maintenance details" continues to be captured in updated PM checklists and PM procedures, and in newly created 1-Point Lessons
Machine metrics are being significantly improved as a result of these maintenance activities

STEP 4 INFRASTRUCTURE SUPPORT

In order for TPM teams to successfully implement Step 4 activities, the TPM steering committee must provide for the following organizational infrastructure to support these new activities:

1. A PM evaluation process.
2. A PM document change, notification, and training process.
3. Documented failure-analysis processes.
4. An effective failure-analysis training process.
5. A process for capturing and organizing machine failures that teams must resolve, i.e., trouble lists.
6. A measurement and reporting system for making the results of all failure-analysis activities visible to management.
7. Systems for measuring advanced machine performance metrics.
8. Changes in evaluation, ranking, and reward systems to support wanted behaviors and stop rewarding old, unwanted behaviors. For example, reward people who engage in knowledge sharing, such as those who author many useful 1-Point Lessons. Also, reward those who successfully prevent machines from breaking down more than people who only repair broken machines.

STEP 4 DELIVERABLES

Step 4 activities should continually deliver a reduction in the failure rates of equipment. At the same time, technicians should find performing the maintenance work that delivers reduced equipment failures easier to complete than all the repair work that they used to perform to keep the machines running. They should spend far less time performing breakdown repairs and instead spend more time on failure analysis and improvement activities. These activities delivered failure-rate reductions of about 80 percent on Agilent equipment in less than one year.

 STEP 4'S MINDSET CHANGE
Equipment failures can be prevented, despite many maintenance technicians' belief that machines will always break down, no matter what maintenance is performed. This self-fulfilling prophecy has kept most maintenance organizations locked into breakdown maintenance.

8

Step 5: Improve
Machine Productivity

This chapter describes how to implement
TPM Step 5, including:

- Performing lubrication, calibration and adjustment, quality maintenance, and machine-part and condition-of-use analyses to improve maintenance plans and machine performance
- Analyzing productivity losses and making improvements to reduce them
- Using extended and continuous condition monitoring as maintenance tools to prevent machine failures

8

STEP 5 GOALS

In Step 1, activities were performed to restore equipment to "new" condition by making all areas of the equipment clean and free of humanly detectable minor defects.

In Step 2, preventive maintenance (PM) plans were identified, organized, and scheduled.

In Step 3, the goal was to achieve PM implementations consistent with the documented maintenance procedures developed in Step 2. This was achieved when "the machine did not know the difference" as to who was performing the maintenance work.

The goals for TPM Step 4 were to prevent equipment breakdowns using failure analysis and to make PMs faster and easier.

In Step 5, the team:

- Takes even higher-level steps to improve machine performance, using more sophisticated analyses and countermeasures to prevent machine failures
- Recognizes that equipment breakdowns are not the only significant equipment productivity losses; other machine productivity losses also need to be reduced

This chapter presents techniques for making additional reductions in machine failure rates and improvements in machine productivity.

LUBRICATION ANALYSIS

Lubrication analysis is a fundamental aspect of equipment maintenance and is part of almost any machine's basic maintenance plan—like those created in TPM Step 2. However, in Step 5, we wish to elevate the lubrication system to a higher level and attempt to completely optimize the lubrication needs of every moving part in every machine.

Carrying out lubrication maintenance in the most effective way means:

- Creating a lubricant scheduling system for every necessary location—perhaps a visually controlled one
- Having the equipment necessary to carry out proper lubrication
- Training people to do lubrication work properly
- Scheduling lubrication work in tech PMs or in operators' cleaning and inspection standards
- Storing lubricants in a properly organized 5S location

Lubrication in an IC fab is somewhat unusual compared to most manufacturing industries. Moving parts in IC equipment generally move only IC wafers and thus are very lightly loaded machines. Most moving parts are "lubed for life" and require no further lubrication. Furthermore, many types of lubricants are not allowed in the clean room at all because they are possible contaminants to the manufacturing process. Thus, there is little opportunity to transfer lubrication maintenance to operators in an IC fab. Lubrication points are few, are generally located beyond operator-accessible machine locations, and are lubricated so infrequently that they are unsuitable for line items on an operator cleaning and inspection standard. Still, lubrication is vital to machine maintenance, even in lightly loaded IC equipment. Also, many of the support facilities required to keep the IC clean room operating—located outside the clean room—are machines that require significant lubrication, which is scheduled in technician PMs.

 Lubrication analysis is a simple but thorough review of all the moving parts in a piece of equipment. We need to be certain that every moving part in the machine is being optimally lubricated.

Maintenance personnel often overlook lubrication, but it is one of the most critical maintenance items on any machine—even one requiring little lubrication. It is common to assume that lubrication maintenance is such a low-level skill that everyone automatically knows how to do it correctly, but like all important maintenance, lubrication must be completed with precision and attention to detail. Taking the time to prepare and follow technically correct lubrication procedures is more work than it might appear, often requiring considerable investigation with component manufacturers. Note that this level of detail was most likely not completed during Step 2 of TPM, when initial lubrication maintenance plans were created.

There are three steps in the lubrication process:

1. Begin by making a list of all the lubrication points on a machine. Every moving part is a candidate for lubrication maintenance:
   - Identify every moving part in the machine. Are any of these parts "self-lubricating" or "lubed for life"? In other words, do any of these parts *not* require any future lubrication? If no lubrication is required, the moving part can be removed from the list.

     There is no way to add lubricant to lubed-for-life bearings, and these are fairly easy to recognize. However, some moving parts that are self-lubricating are difficult to identify. These may need to be investigated with the component's manufacturer.

     For example, a machine bushing made from a high-tech plastic polymer appeared to be a lubrication point in one of our machines, so we faithfully applied grease to this part for years. We were surprised when the bushing repeatedly failed from wear, so we lubricated it even more often. Unknown to us, the part was self-lubricating. We finally discovered the grease we were adding trapped contaminants and hardened into an abrasive mixture that caused accelerated wear in the parts. When we quit lubricating it, the bushing lasted much longer.
   - If lubrication is required, the lubrication point should remain on the lubrication checklist for the machine. A list of all machine lubrication points should now be completed.
2. Once all lubrication points have been identified, specify the lubrication needs required for each:
   - What lubricant is to be used?
   - How often does it need to be applied?
   - How much is to be applied?
   - How is it to be applied?
     - Is it simply added to the lubrication point? If so, what steps are required to prevent the addition of contaminants, such as cleaning a zerk fitting before attaching a grease gun to it?
     - Does the old lubrication need to be removed by cleaning before new lubricant is added? If so, how is it to be cleaned?
     - What tools are required?
3. After the technical aspects of all the lubrication needs are understood, put a system in place to support these needs, including:

- The lubricants themselves, which should be procured parts that are automatically replaced as they are consumed (as would be any regularly scheduled part that is replaced on a machine).
- A 5S location for each lubricant. Each should always have a labeled place where it can be found and should always be kept there.
- A scheduling system that provides for the proper lubrication interval of every lubrication point:
  - Lubrication maintenance requirements can be included in scheduled technician maintenance checklists, as are most other maintenance chores and part replacements. They might also be included in operators' cleaning and inspection standards.
  - A visual identification system for the lubricants can be created in some suitable situations. Color codes for lubricants and symbol shapes at lubrication points can identify the lubricant type and frequency of lubrication. Such visually controlled systems are employed most often for lubrication points tended by machine operators, rather than technicians; still, visual controls can be very useful for technician lubrication systems as well.
- Training material, in the form of 1-Point Lessons or maintenance procedures, must be prepared so that the proper lubrication procedures are permanently documented.
- Training activities must take place to assure that those lubricating the machine are qualified to do so according to the procedures. This is true whether machine operators or equipment technicians carry out the lubrication work.

One of the most common mistakes in machine lubrication is use of the wrong lubricant. "Grease is grease" is a common misconception. People often lubricate a shaft or bearing with whatever lubricant is handy or seems suitable. An experience in our IC fab taught us how the wrong lubricant can cause serious problems.

One mechanism in a machine contained a linear ball bearing on a hardened shaft that was being lubricated with a very viscous grease. The grease was handy because it was being properly applied to a part near the shaft, so the shaft was simply lubricated at the same time with the same product. However, this grease was entirely unsuitable for this particular mechanism and completely "gummed it up," to the point where the motor driving the parts could not overcome the resistance created by the sticky grease. The machine stopped operating and had to be taken out of production. When this failure was investigated, we discovered that a lubricant with entirely different properties was required for proper operation of this mechanism.

Often, several lubrication materials might be suitable for a given part—a bearing, for example. But maintenance consistency requires that the same lubricant be used every time on that part. Lubricating a part at different times with different lubricants may be less effective than using a single lubricant. Mixed lubricants can sometimes considerably shorten the life of a part by interacting and destroying the necessary lubricant qualities.

Figure 8-1 shows a typical 5S lubrication system used inside Agilent's IC fab. A toolbox in each area of the IC fab contains a drawer where technicians can always find the lubricants specified on their PM checklists. Each lubricant is identified with a part number so that a replacement can be readily pulled from the stockroom when the lubricant level approaches empty. Note how few lubricants are used for our IC equipment in the clean room. Many more lubricants are required for equipment located outside the clean room.

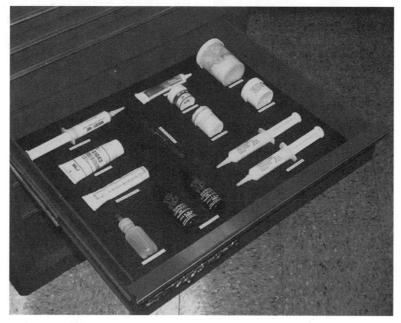

Figure 8-1. This toolbox drawer contains all the lubricants used inside Agilent's clean room

Figure 8-2 shows an effective lubrication system developed by our vacuum pump technicians. Changing oil in vacuum pumps in an IC fab is a large chore because there are so many of them, and process contamination often requires frequent oil changes on numerous pumps. The oil must be removed and properly disposed of, and new lubricant added. Before these new lubrication systems were developed, we removed the pump's drain plug and drained the old oil into a pan. Then, new oil was poured into the filler cap on the pump.

The automatic oil-changing systems developed by our pump techs contain quick-connect fittings that connect to the pumps and allow the oil to be changed quickly, cleanly, and automatically. Because different pumps use different lubricants,

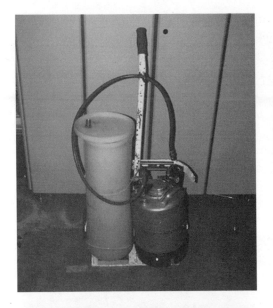

Quick-connect fitting
on pump

Figure 8-2. The two types of vacuum pump oil change carts are shown in the upper photos. Each contains a supply of new, clean oil and a tank to receive the old oil from the pump. Every wet vacuum pump has a quick-connect fitting so that the cart can be easily connected to the pump. The old oil is automatically removed from the pump and replaced with new oil. Each cart uses a different kind of oil and quick-connect fitting, so that only the correct oil can be added to a pump.

carts like those shown in Figure 8-2 are used, each containing a different oil. Different quick-connect fittings on the oil service carts and the vacuum pumps assure that each pump always receives the right type of oil.

CALIBRATION AND ADJUSTMENT ANALYSIS

The object of calibration and adjustment analysis is fairly straightforward—to identify every adjustable component that exists on a machine and document its proper calibration or adjustment procedure.

This process is similar to lubrication analysis; the first step is to physically survey the machine to discover every possible adjustment point on it. Most machines require many adjustments, and a checklist of each machine adjustment should be created. It is not too difficult to complete this survey for most machines, although it may take much longer for large, complex machines.

Once all adjustment points are identified, the following questions must be answered for each:

1. What is the proper adjustment procedure for this point?
2. Does this adjustment require any special tools or instruments to complete? Do we possess these tools?
3. Does the adjustment need to be made at regular intervals? If so, is it scheduled on an appropriate PM?
4. How can I tell at any given time if the adjustment is properly set?

An example of calibration and adjustment analysis follows.

In Agilent's IC fab, hot water circulates to a machine from a water heater subassembly. This hot water is critically important to maintaining the machine's proper process conditions. As a process security measure, the machine manufacturer built an adjustable flow switch into the water circulation loop. If the water flow becomes too low or stops flowing, the flow switch trips and shuts down the machine to protect any wafers in it from being processed in an improper environment.

Unfortunately, our technicians were not aware of the proper method of adjusting this flow switch. When it would trip and shut down a machine, they did not know if the water flow was really too low, or if the flow switch was just improperly adjusted. When a shutdown occurred, most technicians simply adjusted the flow switch until the machine started running again—a shortcut often taken because of pressure to return the machine to production. We didn't believe we had time at that point to learn how to properly adjust this switch. Unfortunately, this situation led to the maladjustment of many flow switches, and many machines were receiving too little heating water flow. Our maladjustment of the flow switches had essentially overridden the manufacturer's safeguard against low water flow.

Our solution to this problem was to obtain the correct adjustment procedure for the flow switches from the manufacturer. We then documented it in our maintenance procedures and included periodic adjustments in our PMs. The flow switches were then correctly set. Those that were "tripped" after being properly adjusted—indicating too little hot water flow—were indicating actual problems. Rather than improperly adjusting the flow switch to get the machine back in operation, our techs traced the cause of the low water flow and repaired that problem.

In almost all cases, the primary cause was impeller wear on the hot water circulation pumps. So the technicians also established a condition-based PM to replace the impellers in the pumps at an appropriate time. Maintaining the pumps prevented low hot water flow, and this—combined with the proper flow switch adjustment—eliminated machine shutdowns from the hot water system.

QUALITY MAINTENANCE ANALYSIS

> *People don't actually maintain quality, they maintain machine hardware. The condition of the machine hardware determines the quality of the manu-factured product.*

Quality maintenance analysis reveals the relationships between desirable quality metrics of the product and certain components in the machine. Maintenance plans are then created to keep these components in their optimal state.

TPM quality maintenance is somewhat different from statistical process control, where the output of the process—the product—is measured, and the measurements are used to adjust and control the process inputs. When control is lost, troubleshooting is conducted to find the cause of the quality problem. The focus of TPM quality maintenance is to always keep certain inputs to the manufacturing process—namely, the machine components—in their optimal state with maintenance plans, independent of product measurements. TPM methods take a proactive approach to maintaining machine components—to keep them "as they should be" without waiting for quality failures to trigger a search for machine component abnormalities.

Table 8-1 is an example of a quality analysis performed on one of Agilent's machines. The machine process required four quality parameters for the finished product: thickness, uniformity, stress, and particles. Engineers and machine technicians traced each quality metric to all hardware that had a strong influence on them.

Table 8-1

Primary Quality Metric	1st Level Controlling Parameter	2nd Level Controlling Parameter	Controlling Hardware
Thickness	Deposition rate	Chemical concentration	LFC
			Helium MFC
			Injector
			Helium regulator
			Heat tape
		Temperature	Susceptor TC
			Heater lamps
			HX subsystem
		Pressure	Baratron
			Throttle valve
			Vacuum pump
			Pump ballasts
		O_2 concentration	O_2 regulator
			O_2 MFC
		RF Power	RF generator
			RF matchwork
			Ground strap
		Gap	Susceptor
		Chamber Age	Etch chamber
			CVD chamber
Uniformity	Gap		Susceptor
	Wafer alignment		Susceptor
	Susceptor age		Susceptor
	O_2 concentration		(See thickness)
	Other hardware		Gas box
			Shower head
			Blocker plate
			Pumping plate
			Mixing block
Stress	Thickness		These 1st-level controlling parameters are already listed above
	RF power		
	Uniformity		
	Chemical concentration		
	Susceptor age		

continued

Table 8-1

Primary Quality Metric	1st Level Controlling Parameter	2nd Level Controlling Parameter	Controlling Hardware
Particles	Wafer handling		Slit valves
			Robot
			Elevator
			Cassette assembly
			CVD lifts
			Etch lifts
			Pneumatic valves
	Inter-wafer cleaning process		End point detector
			Cleaning gas regulators
			Cleaning gas MFCs
	Other hardware		Gas filters
			Chamber O-rings
			Shower head
			Susceptor
			Chamber
			Etch process kit

Once the quality components (those listed in the right-hand column) were identified, maintenance plans were prepared for each one. The goal of these maintenance plans was to keep the quality components in their optimal condition, performing day after day without variation. The theory is, if all quality components are identified and their performance kept consistent, then the product quality produced by this machine should be consistent as well.

> *When creating maintenance plans for quality components, consider not only the restoration of deterioration but also the proper adjustment and calibration of these components. Maintenance plans should be designed to consider, "How do I know for certain that this component is performing as it should be?"*

Table 8-2 summarizes all the quality components identified for this machine and references the maintenance plan that was created to maintain each component's optimal state.[1]

---

[1]Note that the details of the maintenance plans referenced are not shown. The scheduling intervals are described by time in the table, but were actually scheduled by machine usage in our IC plant.

Table 8-2

Quality Component	Maintenance Plan
LFC	Calibrate monthly
Helium MFC	Calibrate monthly
Injector	Clean biannually
Helium regulator	Inspect semiannually
Heat tape	Inspect semiannually
Susceptor TC	Replaced with every new susceptor
Heater lamps	Inspect monthly
HX	Inspect biweekly
Baratron	Zero monthly/calibrate semiannually
Throttle valve	Rebuild quarterly
Vacuum pump	Inspect biweekly/service annually
Pump ballasts	Inspect biweekly
O_2 regulator	Inspect semiannually
O_2 MFC	Inspect/calibrate monthly
RF generator	Inspect/calibrate quarterly
RF matchwork	Inspect/calibrate quarterly
Ground strap	Inspect every susceptor change
Susceptor	Replace bimonthly
Etch chamber	Rebuild semiannually
CVD chamber	Clean/inspect monthly
Gas box	Clean/rebuild biannually
Shower head	Replace bimonthly
Blocker plate	Clean monthly/replace annually
Pumping plate	Inspect monthly
Mixing block	Clean biannually
Slit valves	Rebuild biannually
Robot	Rebuild biannually
Elevator	Inspect and lube quarterly
Cassette assembly	Inspect and lube quarterly
CVD lifts	Rebuild biannually
Etch lifts	Rebuild biannually
Pneumatic valves	Inspect quarterly
End point detector	Inspected and calibrated with Wet-Wipe PM
Cleaning gases regulators	Inspect semiannually
Cleaning gases MFCs	Calibrate monthly
Gas filters	Replace biannually
Chamber O-rings	Replace on various schedules

Consistently maintained machine components provide consistent product quality.

MACHINE-PART ANALYSIS

Machine-part analysis involves discovering all the parts on a machine that require scheduled maintenance. Agilent technicians discovered that many parts in our stockrooms were being used only to replace parts during breakdown maintenance. Our scheduled maintenance plans never called for these parts to be replaced; they had been overlooked in most cases.

Machine-part analysis is undertaken to assure that no parts that should be included in maintenance plans have been accidentally omitted. These parts can be identified by the following criteria.

- *Parts that have been replaced.* Any part that is stocked or has a purchase history. These parts have a known history of requiring replacement and should almost certainly be included in the machine's maintenance plan.
- *Parts that have not been replaced, but probably will be some day.* Any parts that move or dissipate energy and are therefore likely to deteriorate with use. These parts should be considered for inclusion in the machine's maintenance plan. Perhaps they have a long life and have not failed yet, but will in the future.
- *Low-priority parts.* Parts that are expected never to fail, or whose failure is of little or no significance. Maintenance plans are generally not required for these parts.

Once parts that require maintenance plans have been identified, the maintenance planning system described in Step 2, Figure 5-7, can be used to design maintenance plans for these parts.

CONDITION-OF-USE AND LIFE ANALYSIS

Many machine subassemblies may appear to be properly maintained because they are durable. For example, a vacuum pump may last for two or three years before failing. This failure is often assumed to be a normal end-of-life experience. But why didn't the pump last for five years, or even ten? One way of exploring these questions is to identify all the conditions-of-use required for the component to live a natural lifetime. Then the actual machine conditions can be compared to the optimal ones, and any detected abnormalities corrected. Deficiencies in required conditions-of-use cause accelerated deterioration in machine parts.

One of the principal goals of all maintenance work is to keep every part in a machine "as it should be—free of any minor defects." Conditions that are not "as they should be" often cause accelerated deterioration in other parts. This increases machine downtime and the cost of replacement parts, and wastes technician time. Machine deterioration, inadequate machine design, or misuse of the machine can cause abnormalities.

A simple example of condition-of-use analysis follows, considering the life of an automobile tire. Suppose certain radial tires are sold with a life expectancy of 80,000 miles. If the tires last 80,000 miles, then the maintenance plan that was used for them must have been adequate. But what if the tires last only 50,000 miles? Perhaps they were defective, or the marketing was deceptive, but more than likely the tires were not provided the proper conditions-of-use to achieve their 80,000-mile design lifetime. In other words, the tires were subjected to conditions for which they were not designed and which reduced their design lifetime.

When considering all of the conditions that must be maintained for a natural part life, consider the conditions from all of the "4M" sources:

- Man
- Machine
- Material
- Methods

The following list identifies some of the conditions-of-use required to achieve the design life of an automobile tire:

- Tire inflation is always maintained correctly
- Tires are periodically rotated according to a rotation schedule
- Tires are always kept properly balanced
- Wheels on which the tires are mounted have no abnormalities:
  - The rims contain no dents or rust
  - The wheels roll straight—that is, without wobbling out of a plane
- The four wheels on the car are kept aligned with one another
- The brakes contain no performance abnormalities, such as "peddle pulsing" or "grabbing"
- The gross vehicle weight (GVW) of the automobile is suitable for the tire rating—that is, the tire is not placed on too large or too heavy a car, and the car is not overloaded

- The car's suspension does not contain worn parts:
  - Shock absorbers
  - Struts
  - Ball joints
  - Rod ends
  - Steering box
- The designed operating conditions of the car are nominally observed:
  - No "heavy-footed" acceleration or braking
  - No "high-G" cornering
  - The car is driven on smooth pavement
  - The car is driven within legal speed limits on high-speed roadways

The more these required conditions-of-use are violated, and the longer the term of the violations, the shorter the life expectancy of the tires.

Maintaining all the conditions-of-use requires more than maintaining the tires themselves; notice that only the first three conditions stated on the above list refer directly to maintenance actually performed on the tires. The remaining conditions require appropriate maintenance plans for other parts of the automobile and require certain proper behaviors by the operator of the machine—the car's driver.

Parts that fail before their time on production machines may live short lives for the same reasons as the short life of the tire in the above example—their necessary conditions-of-use were not maintained:

1. Proper maintenance was not done on a failing part.
2. Proper maintenance was not done on other machine parts, which induced a higher rate of deterioration in the failing part.
3. The machine or machine component was not used in the manner required for maximum part life.

For example, a condition-of-use analysis was performed by an Agilent TPM team on a number of similar vacuum pumps that historically had about a three- to four-year life—long enough so that lessons from actual failures were few and far between. The condition-of-use analysis was used to determine which conditions had to be maintained to ensure a normal pump life—whatever that might be. At the time they were studied, the pumps were exhibiting no signs of distress. For all we knew, they were living natural lives—we had no way of knowing the potential pump life in this particular situation. The pump manufacturer said that expected life was one to ten years, depending on the application of the pump.

The condition-of-use analysis performed on the pump revealed the following violations:

- Specified cooling-water pressure was not maintained at the pump
- Proper adjustments to the pump's thermostat were not being maintained, creating non-optimal pump operating temperatures

The IC fab contained a cooling water system for its many machines. Thousands of feet of pipe carried cooling water to hundreds of locations. The system contained both a "supply" pipe and a "return" pipe, so water could flow into and out of each machine on the cooling loop. The cooling-water pressure across the machine was the difference between the pressure in the supply pipe and the pressure in the return pipe.

As one can imagine, the pressure in the supply pipe diminished as it traveled through the plant, getting further and further away from the pump that circulated water through the cooling loop. Likewise, the pressure in the return pipe also diminished as it approached the pump. Therefore, the pressure across the supply and return pipes varied considerably, depending on its location within the plant. Near the pump, the pressure difference was about 70 psi, the supply pressure being 80 psi and the return pressure being 10 psi. But at certain points in the cooling-water piping, the differential pressure was found to be as little as 35 psi. This was barely enough to meet the pressure requirements of many of our vacuum pumps, which required a minimum cooling water pressure of 30 psi.

Pumps were connected to the cooling-water piping system by flexible hoses. Most of these hoses were 1/4" diameter, and where the pressure in the cooling loop was sufficiently high, this was an adequate tubing diameter to supply the cooling water pressures required at the pump. However, in locations on the cooling-water loop where differential pressures were low, the 1/4" flex hose created so much additional pressure loss that the actual cooling-water pressure at the pump was only about 15 psi, even though gauges on the cooling loop showed 35 psi available a few feet away. In those situations, the 1/4" tubing was replaced with 3/8" tubing so that 30 psi pressure was available at the pump, as prescribed by the manufacturer. This situation is illustrated in Figure 8-3.

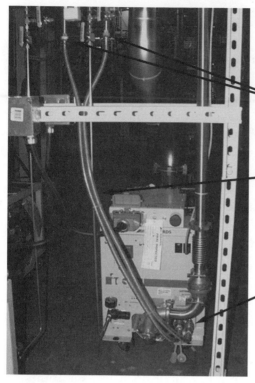

Differential pressure of the cooling water is adequate at the supply and return sources

Flexible cooling lines connect the pump to the cooling supply loop

Differential pressure of the cooling water at the pump was inadequate before the flexible tubing size was increased

Figure 8-3. Flexible hoses connecting a vacuum pump to its cooling water supply

Vacuum pump life also depended on the pump's operating temperature, which needed to be regulated so that it was neither too hot nor too cold. Either extreme caused increased rates of deterioration. The pumps had two cooling mechanisms: cooling fins on some parts of the pump, and cooling water circulating through other parts.

The cooling fins needed to be maintained contamination-free to maximize their heat transfer rate. This was accomplished in our first TPM step—creating clean and defect-free machines. However, when actual pump temperatures were measured, many pumps were found to be running either too hot or too cold. The thermostats that controlled pump temperature were adjustable but had always been left at their factory settings. However, the pumps were used on many different kinds of machines and operated at many different points within their range of capability, so adjustments had to be made in some of the thermostat settings in order to maintain the proper temperature in each pump.

Figure 8-4 shows a vacuum pump having its temperature checked to make certain its thermostat is properly adjusted.

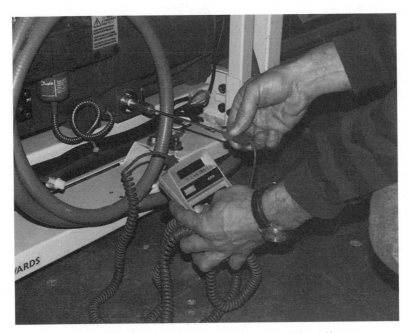

Figure 8-4. Measuring pump temperature to maintain optimum
operating conditions

In yet another situation, a small cooling-water circuit was designed to supply
125 gpm of cooling water to a group of machines. The pump curve shown in Figure 8-5 shows the performance of the pump selected to circulate the water.

As can be seen from the performance curve, the pump was available with
three different motor sizes, depending on the design operating point. The more
closely the motor was matched to the actual pump power need, the more efficient
it was. Therefore, a 5-horsepower motor was placed on this pump when the original cooling circuit was installed. However, over the years, more and more
machines were added to the cooling loop, which caused the operating point of the
pump to shift. The new operating point of the pump drew about 5.85 horsepower. The pump motor had a 1.15 service factor rating, which meant it could
produce 15 percent more than its 5-horsepower rating, or about 5.75 horsepower. This meant the pump motor was now working beyond the extreme limits
of its capabilities. This situation remained unnoticed because the motor continued to run, and no obvious clues to improper motor conditions-of-use were
noticeable.

However, once a condition-of-use analysis was undertaken, it uncovered the
high motor-load situation. The 5-horsepower motor was now very likely to fail
long before its time. To resolve the problem, the 5-horsepower motor was
replaced with a 7.5-horsepower motor.

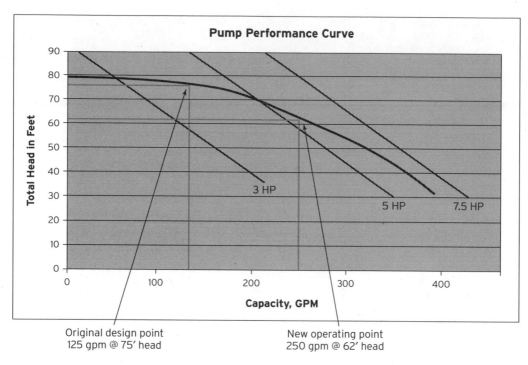

Pump Performance Curve

Original design point
125 gpm @ 75' head

New operating point
250 gpm @ 62' head

Figure 8-5. Pump performance curve indicating how changing conditions caused shorter motor life

All these improvements came from a condition-of-use analysis. No failure analyses of failed equipment was necessary—we improved our maintenance plans and extended equipment life simply by learning more about the conditions needed to assure that parts lasted as long as possible.

PRODUCTIVITY ANALYSIS: THE SIX BIG PRODUCTIVITY LOSSES

Much of the TPM work accomplished so far has focused on preventing equipment breakdowns. However, there are many productivity losses on the manufacturing floor and in the machines other than equipment breakdowns.

Equipment losses are most often described as the "six big losses." In Step 5 of TPM, the team explores the true nature of productivity losses.

 Some losses are visible and obvious to people—for instance, a scrapped product. But many productivity losses are hidden from view. They are either difficult to measure, or they are visible but people do not perceive them as losses.

Although we speak of the "six big losses," sixteen major productivity losses have actually been identified for most factory floors. These include management

losses, operating motion losses, line organization losses, and energy losses. However, in Step 5, the six most common machine productivity losses—defined for a process operation like an IC fab—are addressed:

- Machine breakdowns
- Machine setup and adjustment time
- Product scrap
- Low product yields
- Minor machine stoppages
- Reduced machine speed

The following section describes these losses in more detail and helps teams identify them on machines, so they can be reduced or eliminated.

Machine Breakdowns

Machine availability is one of the most readily identified and easily measured machine productivity losses. Any time that a machine scheduled to run in production cannot do so because of a breakdown, productivity is lost. People easily perceive machine breakdowns to be productivity losses.

However, breakdowns on different machines can have different effects on overall factory productivity. Any hour lost on a bottleneck machine is an hour of factory throughput lost forever. But lost time on a non-bottleneck machine can affect total factory productivity in different ways. A non-bottleneck machine that fails for one hour a day every day for 90 days is available 96 percent of the time, and, given enough excess capacity, this daily loss might be of little relevance to the flow of material through the factory bottleneck. But a machine that breaks down only once every three months and stays down for ninety hours also has a 96 percent availability. However, a machine failure lasting four days would certainly hurt the throughput of any manufacturing line.

 Machine breakdowns—as a performance metric—need to have more than their long-term averages improved. The day-to-day (or minute-to-minute) reliability of any machine's availability must also be improved as well.

Machine Setup and Adjustment Time

Another productivity loss that can be measured on most equipment with relative ease is setup time. This is the time consumed on a production machine being prepared for processing the next product or batch, but not yet actually processing the parts. Setup might involve tooling changes, process recipe changes, machine adjustments, or quality tests.

Many people don't perceive these setup activities as productivity losses, because they are a "normal and necessary" part of the machine operation. However, setup causes productivity losses because these operations could theoretically be completed in a reduced amount of time, or ways could be found to eliminate them altogether.

Once setup time is recognized as a loss and measured, steps can often be taken to reduce setup losses. The following example is a project undertaken at Agilent to reduce the setup losses of our bottleneck machines.

Product lots moved through the bottleneck machine set, as illustrated in Figure 8-6. (Agilent's bottleneck "machine" was actually three machines mechanically linked together and controlled as if they were a single machine.)

This arrangement had a major flaw—the machine set could only process one lot at a time because the machine set controller could only hold one process

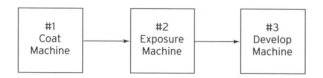

Figure 8-6. These three machines, mechanically linked together, were controlled by a single machine controller and behaved as if they were a single machine.

recipe at a time for the entire set of three machines. Since each lot usually required a different process recipe for each machine, an entire production lot of wafers had to completely exit the third machine before a new lot could be started into the first machine. This meant hours of idle time on each machine, amounting to a setup loss of about 20 percent of each machine's operating time.

To reduce this setup loss, our engineers first added new capabilities to the machine set controller. The new controller could operate the three machines more independently than before. Once they accomplished this, however, they had a new problem on their hands. Since there might now be as many as three production lots in this machine at any one time, they had to devise a system that operators could use to keep track of these lots without any possibility of confusing them or running the wrong process recipe. They solved this problem by providing the operators with an organized, visual interface—a 5S-type of machine-input table to keep track of production lots that were either in the machines or preparing to enter them. They also provided automatic communication between this table and the machine controller with automatic bar-code readers.

This system allowed production lots to continually flow into the first machine, even while the second and third machines were still processing other lots with different process recipes. It also provided a continuous flow of wafers through all of the machines. The bar-code readers on the table automatically downloaded the correct process recipe into each of the appropriate machines as the first wafer of the lot moved into that machine.

This visually controlled table, and the software operating it, reduced the set-up loss of this machine set to near zero, resulting in a 20 percent increase in capacity of the plant's bottleneck machines. Figure 8-7 shows the 5S automated table that helped make this improvement workable in the IC fab. It also illustrates how automation and visually controlled manual operations can be integrated with one another.

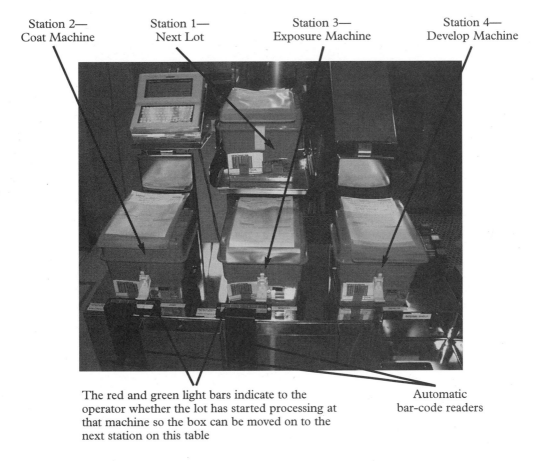

Station 2— Coat Machine

Station 1— Next Lot

Station 3— Exposure Machine

Station 4— Develop Machine

The red and green light bars indicate to the operator whether the lot has started processing at that machine so the box can be moved on to the next station on this table

Automatic bar-code readers

Figure 8-7. A visually controlled "input station" feeding lots into a series of linked machines

Product Scrap

Product scrap, another relatively obvious productivity loss, is not only a loss of the material included in the product, but also represents a loss of all the time that all the machines on the production line "wasted" producing this scrap product. Sometimes, poor-quality product can be salvaged by rework operations but these, too, represent lost productivity. The bottleneck machine loss is especially intolerable, as it can never be recovered. Scrap products in a factory running at full capacity are often lost sales.

However, not all off-quality product is measured as scrap on the manufacturing line, even though it should be. Any poor-quality product that is shipped to customers represents an even larger productivity loss for the factory than scrap on the production floor. This is because of the high cost of retrieving and replacing the poor product. It is far cheaper to scrap a poor product on the production line than it is to send it through the distribution system into a customer's hands. The expense of removing the poor product from the field and replacing it with a good product might even include the loss of a customer to competitors.

 Sending poor-quality product down the production line to artificially improve scrap measurements is an extremely foolish manufacturing process.

Low Product Yields

In a process factory like Agilent's IC fab, not all scrap product on the manufacturing floor is obvious. A broken or misprocessed wafer is easily recognized as scrap. But sometimes machines seem to be making good-quality product when actually they are operating at some level of diminished quality.

An IC wafer can have dozens, hundreds, or even thousands of chips on it. The chip is the real product that our customers purchase. Any manufacturing process that reduces the yield of good chips on our wafers causes a productivity loss. Unfortunately, this productivity loss is not obvious to people on the factory floor at the time it actually occurs. Instead, it is detected at the end of the manufacturing process, when the chips are sawed from the wafer. The good chips are packaged and sold as finished goods from our factory, but the defective chips are scrapped.

Minor Machine Stoppages

Machines often stop on the factory floor because of a product jam, a machine error, or a warning given to an operator. In these situations, operators usually can restore the machine to normal operation themselves with some minor action as soon as they notice that the machine has stopped. In most factories, minor stoppages are considered only a minor productivity loss because they are so short in duration; they therefore are not recorded on any kind of machine failure log. However, even short-duration stoppages—when they occur often—can add up to a significant productivity loss. In many factories, minor stoppages are a major contribution to lost productivity. But they are often overlooked because the education needed to recognize the significance of these losses, or a system to measure them, is lacking.

Reduced Machine Speed

Reduced machine speed is another significant productivity loss that is often over-looked because it is so difficult to recognize. Speed losses have several causes, as described below.

- Sometimes machines slow down from their normal operating speed because of abnormalities in their hardware. As long as the machine is running, however, the slow speed often goes unrecognized. Agilent once had a machine take eight hours to process a batch of wafers it usually completed in two hours. But without any warning signals from the machine, our operators—who operate many machines simultaneously—did not notice its slow pace. It was caused by a cryopump that was running slowly because of contamination buildup.

- Sometimes, maintenance technicians slow down machines that are misbehaving because this is a quick way to keep them from producing errors or problems. Taking the time to find and correct the true root cause of the problem would almost certainly be less of a productivity loss than slowing down the machine—although it is a less expedient repair. However, a machine in this situation may need to have its speed decreased further in the near future as the true root cause of problems continues to worsen.

- Other speed losses stem from the machine's control system. For example, IC processing machines usually carry out multiple operations on wafers, so they contain many different process recipes. If the engineer who designed them included excess time or unnecessary pauses in the recipes, they will run slower than they need to. If a recipe takes longer to process a wafer than it could if some of its parameters were modified, the machine will also run slower than is theoretically possible. Most people accept these invisible losses as normal, but they are clearly productivity losses.

    One such speed loss was found on an Agilent machine that deposits material onto wafers in a vacuum chamber by mixing gases in a plasma above the wafer. This deposition is not confined to the wafer—it is actually deposited throughout the inside of the process chamber. Once the wafer receives the proper thickness of material, it is removed from the chamber. But before the next wafer can enter this chamber, the material that was deposited on the chamber's parts during the previous wafer's production must be cleaned off. If this material is not removed after each wafer deposition, it builds up on the parts in the process chamber and begins to interfere with moving parts inside the chamber. It also creates a particle storm inside the chamber, which contaminates the wafers being processed. Therefore, after a chamber deposition process, the chamber goes through a cleaning cycle to remove the material deposited on it.

In Agilent's situation, this cleaning process took as long as the wafer deposition process. In other words, half of all the time that this machine was "running" product at full speed, it was really not performing any useful work on the product at all—it was just cleaning itself. Again, most people did not consider this cleaning time as a machine productivity loss because it was a necessary part of this machine's manufacturing process. But when the process engineer recognized the productivity loss, he redesigned the cleaning process recipe so that it would be completed more rapidly. In this situation, a mere 25 percent reduction in cleaning time produced nearly a 25 percent increase in wafer capacity of this machine—a significant increase for a situation that most people had not even thought of as a productivity loss.

- Other speed losses stem from the machine's basic operation. The machine may carry out many operations in series, when some of them could be done in parallel. This lost productivity is often recognized as a design limit because that is the way the machine normally runs. (Machines operating at their "normal" speeds are not considered to have speed losses.) Agilent engineers have often done a "good, old-fashioned, stopwatch time study" on a machine to watch how it operates. When they plot the critical path of what really needs to happen in the machine, they often discover many opportunities for speeding it up.

Until they are first recognized as such, productivity losses in machines cannot be reduced or eliminated. Agilent has spent a great deal of effort during TPM Step 5 teaching operators, techs, engineers, and managers what machine productivity loss really is—starting with the "six big equipment losses." Once people recognize productivity losses, they can undertake activities in this TPM step to begin to reduce them.

EXTENDED CONDITION MONITORING

In earlier TPM steps, machine conditions were often detected primarily by human senses (except taste). In Step 5, specialized maintenance technologies are employed that allow technicians to determine machine conditions beyond what human senses and their current instruments can discern. If a machine part is deteriorating gracefully, but the deterioration cannot be monitored by human senses, then one of the following common condition-monitoring technologies might be used to expose the deterioration. (Other technologies are also available in addition to these few examples.)

- Vibration analysis
- Ultrasonic analysis

- Wear particle (lubricant) analysis
- Infrared thermography
- Video imaging
- Water-quality analysis
- Motor-condition analysis
- Jigs, fixtures, and test gauges

Vibration Analysis

Vibration analysis is a highly advanced machine-condition monitoring technology commonly used today in many factory maintenance departments. Vibration-analysis tools are built by many different manufacturers and are used to measure deterioration in rotating machinery. Figure 8-8 shows bearings in a motor being monitored by a vibration-analysis tool. The vibration patterns of the bearings are checked at various intervals and recorded. As the bearings deteriorate, the vibration pattern changes in interpretable ways, even to the point of providing insight into whether a bearing is failing on the inner or outer race, or if the rollers are failing. Vibration analysis usually provides ample warning of impending failure, so techs can plan ahead to replace a bearing or other part.

Vibration analysis on machines can also determine abnormal conditions such as imbalance in rotating parts, looseness, cracks, "soft-footed" mountings, misaligned parts, and even worn keys and keyways.

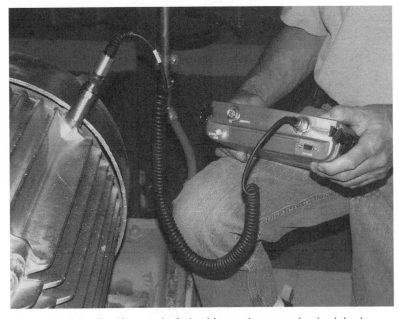

Figure 8-8. A vibration-analysis tool is used on a motor to detect bearing deterioration

Ultrasonic Analysis

Ultrasonic-analysis tools monitor high-frequency machine vibrations—somewhat like vibration-analysis tools—and also "sound" waves produced by machines that have frequencies beyond the range of human hearing. Ultrasonic sounds can be detected and reduced to frequencies audible to people, or they can be displayed on a meter for people to see. Changes in ultrasonic patterns in machines are used to detect cracked or loose parts. They are also used to "listen" to small leaks in lines containing air or other gases. The line might be pressurized and leaking "out," or the line may be under vacuum and leaking "in." Either way, such small leaks are often very difficult to detect. An ultrasonic listening device can detect the ultrasonic "noise" produced by these leaks. This tool has helped Agilent technicians check machines for very minor leaks and correct them before they got worse. Small leaks, like other minor machine defects, can contribute to machine failures. Figure 8-9 shows an ultrasonic listening device being used to test a machine for minor leaks.

Figure 8-9. An ultrasonic leak detector in use, searching for minor gas line leaks

Wear Particle Analysis

Wear particle (lubricant) analysis is a technique usually used for analyzing oil from a machine such as an engine, refrigeration compressor, or wet vacuum pump. Contaminates found in the oil can lead to early detection of machine deterioration. For instance, Agilent uses wear particle analysis on its large refrigeration compressors to detect metal particles caused by abnormal wear before the damage becomes so great that the chiller compressor fails. The analysis allows us to detect minor abnormalities in the equipment that our human senses cannot detect. Discovering these abnormalities early—while they are still minor problems—allows us to take corrective action early on, instead of in the panic of an emergency repair of a down chiller. Figure 8-10 shows a chiller compressor at Agilent that consistently undergoes wear particle analysis.

Figure 8-10. An Agilent chiller, which is inspected routinely for minor defects with wear particle analysis

Infrared Thermography

Infrared-thermography tools translate infrared frequencies to visible frequencies, making heat that our eyes cannot see visible. Agilent uses this technology to detect abnormal heat patterns, which can reveal minor abnormalities in the equipment. Our electricians also use this technology to determine when electrical connections are becoming loose; when even slightly loose, these connections start to produce heat. It would be imprudent to try to sense this small level of heat with a human hand because the terminals being measured are carrying high voltages.

Infrared thermography allows every electrical connection in the power-distribution system in our factory to be routinely inspected. Connections that are beginning to show minor signs of heating are placed on a PM list for tightening. This allows us to power down the equipment and tighten the connections when it is most convenient for us, before a defect grows from minor to major. Figure 8-11 shows the results of an infrared-thermography tool being used to inspect electrical connections.

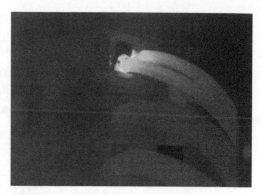

Figure 8-11. An infrared-thermography tool makes visible (right photo) the loose connection on the circuit breaker shown at the left. (Photos courtesy of Electro-Test, Inc., Denver, Colorado.)

Video-Imaging Analysis

Video-imaging analysis tools can provide either high-speed or low-speed analysis of machine motions:

- Consistent problems involving rapid machine operations can be recorded at high speed and then shown at slow speed. This allows a team to detect problems that the human eye otherwise cannot discern.
- Sometimes, video can help detect problems that occur too infrequently for humans to reliably detect; for example, an intermittent failure that occurs only once every few days can be captured on long-playing video so that conditions associated with the failure can be discovered.

Agilent technicians have frequently employed these techniques, using a contractor who owned the necessary camera equipment. In several cases, these tools made available to us information valuable to failure analysis and countermeasure design.

Water-Quality Analysis

Water-quality analysis tools are used on many of the closed water-circulation loops in Agilent's IC fab to detect and correct minor problems with water quality before major damage is done. Agilent's facility contains many closed water-cooling loops used for machine temperature control, and hundreds of branches on this loop contain flow meters, allowing technicians to measure the flow to any piece of equipment. However, if the quality of the water deteriorates—for example, from bacterial growth—the internal surfaces of the piping system can quickly become covered with a slimy coating of material, which reduces heat transfer, increases pumping power requirements, and clouds flow meters so that they are impossible to read. The only way to restore these flow meters to useable condition is to remove them and manually clean them. This requires significant technician time and machine downtime.

To analyze the problem, Agilent began performing regular water-quality analysis on samples of water drawn from all closed water-circulating loops. Minor abnormalities in the water were detected with tools that measured beyond the range of human senses, and minor defects were corrected by water treatment before major damage was done to the system.

Motor-Condition Analysis

Motor-condition analysis tools have a wide variety of capabilities and are available from many manufacturers. Motors can be tested for load, torque, RPM, efficiency, service factor, power balance, winding insulation qualities, over-currents, delaminating stators, bearing condition, and other parameters, depending on the motor test equipment being used. Some of these tests require the motor to be off-line. Others can be done while the motor is in normal production service. Either way, the motor data are recorded and compared to previous readings. Deterioration in any motor condition is made readily apparent by the regular use of these condition analyzers. Here again, instruments that extend the range of our human senses are used to detect small equipment abnormalities so that problems can be corrected early on, at our convenience, instead of in a down-machine situation. Figure 8-12 shows a motor-condition analysis tool in use.

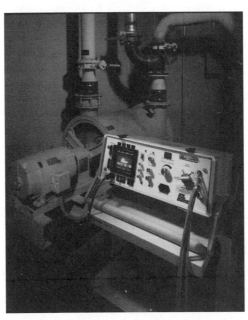

Figure 8-12. A motor-condition analyzer is used to check a motor's winding insulation properties. (Photo Courtesy of Baker Instrument Co., Fort Collins, Colorado.)

Jigs, Fixtures, and Test Gauges

Jigs, fixtures, and test gauges are often useful for extended condition monitoring. For example, Agilent wafers are handled in plastic cassettes that hold 25 wafers. When placed in a machine, the cassette becomes part of the machine. An abnormal cassette creates a temporary abnormality in the machine, which most often causes intermittent wafer-handling problems. Like any machine part, the cassettes must be kept "as they should be," without any minor defects. As the cassettes age, they tend to warp slightly and then won't sit flat on the cassette station. This slight warpage in the cassettes is humanly detectable, but the standard is subjective; people have trouble determining when it becomes a problem.

A simple test tool was purchased that performs cassette warpage inspections with far more accuracy than a person is capable of achieving. This inspection tool is shown in Figure 8-13.

Push handle to insert probes

PASS/FAIL lights

Figure 8-13. A simple cassette-testing fixture that extends human senses to find minor abnormalities in wafer cassettes. The green and red lights indicate "Pass" or "Fail" when the probe is inserted into the cassette.

Expensive equipment for extended condition monitoring should not be used until a need for it has been identified. Throwing money into expensive but unnecessary tools will not enhance TPM activities. Also, consider whether to purchase condition-monitoring equipment or subcontract the work to an outside group. If purchasing the equipment makes sense, develop a thorough training program for the maintenance people who will be using it. Such tools are useless if no one knows how to operate them properly.

Continuous Condition Monitoring

Continuous condition-monitoring tools are sometimes required to resolve maintenance and machine design dilemmas. The maintenance development process[2] sometimes leads to a machine redesign, which might consist of adding a continuous condition-monitoring device. This technique is useful if failures can be detected immediately with continuous condition-monitoring equipment. The manufacturing process can then be halted, and the machine repaired with less productivity loss to the manufacturing line than would be encountered without such detection. In such a case, immediate failure detection is the next-best option to failure prevention. An example of such a situation in Agilent's IC fab follows.

Reticles are one of the tools required in an IC fab. A reticle is a quartz plate with chrome lines and patterns on it used to pattern the circuits on an IC wafer. Reticles are very expensive to produce, and one is required for every layer of every product that we manufacture. Agilent owns thousands of reticles. Reticles are very fragile and subject to damage from static electricity. A static charge jumping to the chrome on the quartz surface can remove a small amount of the chrome. This changes the pattern on the reticle, which effectively changes the electrical circuit that gets built on the wafer.

Such reticle defects are almost impossible to detect until the wafers complete their manufacturing process and then fail final quality tests. Autopsies on these wafers may reveal the cause of the failure but, as one might suspect, this is extremely slow and difficult work for Agilent engineers. Such a failure can easily produce millions of dollars of scrap product before it can be detected and corrected. Therefore, it is critical for us to prevent reticle damage.

To protect reticles from static charges, they are stored in environmentally controlled storage cabinets. These cabinets contain high-voltage ionizers that continuously maintain a static-free environment within the storage cabinet; without ionizer protection, static charges are likely to occur when an operator reaches into the cabinet to remove a reticle. The ionizers are 100 percent solid-state devices with a theoretically infinite life. They require no preventive maintenance except to maintain their proper conditions-of-use.

However, the ionizers have failed on more than one occasion for various reasons—always unpredictably and unobserved. One circuit board failed which had never failed before, nor has it since.

Ionizer failures in the reticle cabinet are intolerable, yet we have been unable to completely prevent them by any maintenance or redesign means.

In this situation, immediate detection of an ionizer failure would be almost as good as preventing the failure. If an ionizer failure is immediately detected, a reticle

---

[2]This process is illustrated in Figure 5-7.

cabinet warning can be issued and operators can immediately cease their access into the cabinet. If no one touches the reticles, no harm will be done. The challenge is to design a way to continuously monitor the condition of the ionizers.

It was decided that two separate condition monitors were needed. The first monitor detected high voltage at the ionizer discharge tips. If high voltage was present, we were assured that the entire ionizer circuit was working properly. The sensor used to detect high voltage at the ionizer discharge tips was a circuit board from a Christmas tree light tester. This product cost less than $20.00. It is shown in Figure 8-14, installed in a custom-made case.

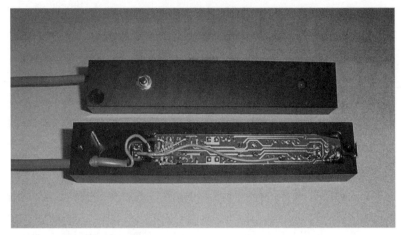

Figure 8-14. An inexpensive, off-the-shelf high-voltage detector circuit from a Christmas tree light tester is used to continuously monitor voltage at the ionizer tips

We also wanted to be absolutely certain that no static charge existed in the reticle cabinet, even though we were certain that the ionizers were working. This sounds like a difficult technical challenge, but it was surprisingly simple for our technicians. This sensor began as a homemade capacitor—two metal washers on a wooden stick. A commercially available static-charge sensor is now used and is shown in Figure 8-15.

These sensors are connected to a PLC, which our technicians programmed to continually monitor the sensors, average their values, and display the measurements on a small screen mounted on the outside of each reticle cabinet. The screen displays the values and allowable ranges for the ionizer voltages and the static charge in the cabinet. Should either one of these conditions go outside of its allowable range, the display flashes a red warning light and produces an audible signal that notifies our operators that the reticle cabinet cannot be safely opened. Maintenance technicians are then called to repair the problem. The display used in this device is shown in Figure 8-16. The display, and the PLC that drives it, cost a total of approximately $500.00.

Figure 8-15. A simple static charge sensor

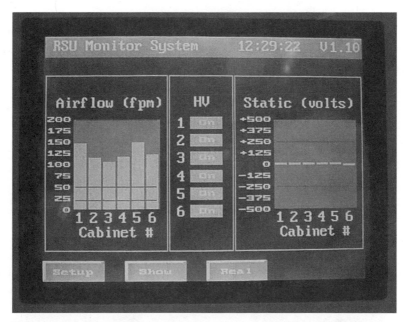

Figure 8-16. The display shows data received from the continuous-monitoring devices in the reticle cabinet on its right screen

With the right technical skills, any machine problem can be diminished, even if it cannot be completely prevented. In this case, the solution was surprisingly simple and inexpensive.

MAINTENANCE COST ANALYSIS

Another simple technique used in TPM Step 5 to detect factory productivity losses is to measure maintenance costs. High costs are often associated with tools that are not running at optimal efficiency. These can provide great opportunities for maintenance cost savings.

To save on maintenance costs, questions first need to be asked about the current situation. How are maintenance man-hours distributed among the equipment? How much money is each machine consuming in spare part replacements? Tracking this information can lead to some insights about machine productivity losses.

One of Agilent's machines—by design—consumes a lot of replacement parts. Yet a cost analysis of the consumable parts being used on this machine surprised us; it revealed that many more parts were being consumed than we had assumed. This would never have been noticed had we not decided to measure the costs of this machine's replacement part consumption. Technicians tracking the use of these high consumables found that, often, they were being replaced prematurely because of some type of mishap. Once damaged in any way, they had to be replaced with new parts.

The unexpected damage had a variety of causes. By focusing on each of these causes and implementing countermeasures, the part consumption dropped considerably, to a level close to what was expected. Because this machine consumed such a high-dollar value of replacement parts, these small improvements paid off in a large way. Making improvements to the lifetimes of these expensive parts was worth more to us than improvements we might have made in extending the life of other, less expensive parts. Figure 8-17 shows the machine and indicates one of the frequently replaced parts.

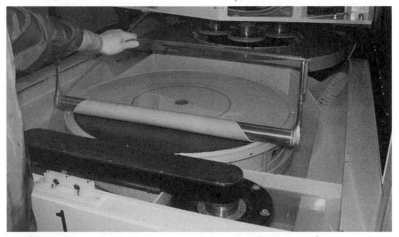

Figure 8-17. An Agilent machine that requires frequent replacement of an expensive lapping pad

STEP 5 MASTER CHECKLIST

As TPM teams implement activities using the Step 5 toolbox on their equipment, they can refer to the following checklist to aid them in pursuing all of the activities of this step. Successfully completing all of the line items on the master checklist completes their Step 5 activities.

STEP 5 MASTER CHECKLIST: IMPROVE MACHINE PRODUCTIVITY

The following methods are being used to further improve machine maintenance plans and equipment designs and to eliminate other machine productivity losses. (Indicate the activities undertaken on each machine and the results achieved by each.)
☐ Lubrication analysis
☐ Calibration and adjustment analysis
☐ Quality maintenance analysis
☐ Machine-part analysis
☐ Condition-of-use and life analysis
☐ Productivity analysis:
  - Availability (downtime and setup losses)
  - Quality (scrap, yield, and rework losses)
  - Speed (minor stoppages and slow operation losses)
☐ Extended condition monitoring
☐ Continuous condition monitoring
☐ Maintenance cost analysis

Technician maintenance roles have elevated from repairing broken machines to activities that prevent machine failures.
Operator maintenance roles have elevated from cleaning and inspection work to higher levels of machine maintenance chores.
Continual learning and improvement activities are a normal routine on the factory floor for all employees
Equipment failure rates have been reduced by about 90 percent
Equipment productivity has been improved by 30 percent or more

STEP 5 INFRASTRUCTURE SUPPORT

In order for TPM teams to carry out Step 5 activities, the TPM steering committee must supply the organizational and infrastructure resources required to support the team activities. These include the following:

1. Documented procedures for each of the tools described in the Step 5 toolbox.
2. Acquisition of the appropriate tools for extended condition monitoring.
3. Training support that continually elevates the knowledge and skill of operators and maintenance technicians.
4. Any resources required by the team that are suitable to pursuing the goals of this step.
5. Changes in evaluation, ranking, and reward systems to support wanted behaviors and stop rewarding old, unwanted behaviors. For example, offer greater rewards to those who contribute to improvement activities than to those who only engage in routine production and maintenance activities.

STEP 5 DELIVERABLES

Completing Step 5 of TPM completes the first TPM level. At this point, in an average factory, machine failure rates should be reduced by 90-95 percent over the failure rates of equipment before TPM began. Machine productivity should have increased by about 30–35 percent. The skill level of the workforce should be raised considerably over that existing before TPM activities began. Continuous learning and continuous factory floor improvement activities should be part of the normal, day-to-day operating culture of the factory. In essence, TPM:

- Reduced equipment failure rates
- Increased factory productivity
- Elevated personnel skills
- Developed a culture of continuous improvement in the factory's operating environment

---

STEP 5'S MINDSET CHANGE
Many people believe equipment productivity losses are caused by the design limits of machines, and that factory personnel cannot affect them. In fact, significant productivity losses can be eliminated by teamwork among production operators, maintenance technicians, and production engineers.

SUMMARY OF TPM STEPS

TPM steps can be applied at both an organizational level and at a team level. For instance, Step 3—achieving precision maintenance—was described earlier in this book from an organizational point of view. Basic technical training, for instance, is provided by management to all maintenance people, and is an activity beyond the scope of any individual team. But every TPM team on the factory floor can also apply all of the TPM steps to their own machines using means that are within their control. The following summary describes the TPM steps from a team's point of view.

Preliminary Step

Use conventional problem-solving means to resolve any severe operating problems the machine might be having before taking a more systematic approach to improving its performance.

Step 1

Remove all humanly detectable minor machine defects from the machine. Once it is clean and defect free, put cleaning and inspection plans in place to keep it that way.

Step 2

Assemble a complete maintenance plan for the machine. Merge the machine vendor's recommended maintenance plan and your own experience with operating and maintaining the machine. Make certain to include areas of the machine that have a history of breakdowns.

Step 3

Carry out your maintenance plan with precision. Make certain that every aspect of the plan is performed identically by everyone involved in it.

Step 4

Prevent machine failures from recurring by applying failure analysis methods to them. Also, continually evaluate maintenance PM's to find ways to make each one "easier, faster, and better."

Step 5

Measure all productivity losses on the machine and systematically eliminate them.

These steps are not limited to serial implementation—they can be done somewhat in parallel. For instance, Steps 1 and 2 can be started at the same time, and Steps 3, 4, and 5 can all be underway at the same time.

Step 5 activities also can be applied globally, beyond the realm of individual machine action teams. This step is most often referred to as the "focused improvement pillar" of TPM. The productivity of machines and manufacturing lines can be measured, and the critical few factory limiters identified. Then, special cross-departmental focus teams can be created to deal with these few critical issues, one team for each issue. These teams supplement the long-standing TPM action team structure. They are created and chartered to resolve a single critical loss, and are disbanded after the issue is resolved.

The focused improvement process is as follows:

1. Collect appropriate data regarding the performance of each machine or machine set on the factory floor.
2. Analyze the data to identify the critical few areas that most limit factory output.
3. For each limiting area, identify the issue that most requires improvement.
4. Create an appropriately staffed focus team to attack each issue identified.
5. Standardize the solutions found across all similar equipment.
6. Continually repeat the preceding steps, keeping appropriate resources focused on the most critical factory losses.

PART 3

Leading the Change

Part 3 addresses TPM implementation—what management must
do to assure successful TPM:

- **Creating a vision, strategy, and tactical plan**

- **Focusing machine improvement activities, including how
  TPM differs from a Pareto approach**

- **Measuring machine performance using various
  improvement metrics**

- **Managing new behaviors using basic performance
  management tools**

- **Sustaining TPM changes—how to make TPM
  into a long-term practice**

- **Team activity boards**

- **Final thoughts**

9

Vision, Strategy, and Tactical Plan

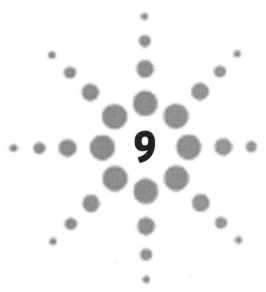

9

A TPM steering committee needs a very clear vision of what it expects its TPM program to accomplish. It must create a strategy, and a tactical plan to accomplish that strategy, before it introduces new practices into the organization. Change is very disruptive, and nonproductive change can actually reduce manufacturing productivity rather than improve it.

Not many production managers are likely to support trading short-term losses for promised long-term gains. Factories must continue to meet today's production goals while preparing to meet tomorrow's needs. That is why every TPM action team must "pay its own way" relatively quickly, providing additional manufacturing capacity with its activities rather than consuming capacity by having machines down for their improvement activities.

Agilent's original TPM pilot team learned this lesson when it first began implementing machine cleaning and inspection activities. Their initial effort went slowly and produced few results to show for long hours of work. This raised skepticism as to whether TPM would, in fact, improve factory productivity. It seemed to be nothing more than another management "project of the month" that only further drew our factory floor people away from their primary task of manufacturing product.

The five-step method for implementing TPM that is described in this work grew from these early team frustrations. Agilent equipment teams were able to improve equipment much more quickly once they began to follow this method. Our activities, as described in this book, are still traditional TPM activities, but we organized our TPM implementation a little differently from the normal procedures.

TPM is a proven process that has been developed in many manufacturing companies over the past 30 years. Companies should not try to re-invent TPM from scratch. Rather, managers responsible for productivity improvement need to comprehend the principles of TPM already developed and apply them to their own needs within their own organizations. They should be extremely clear about their intentions and have a well laid out roadmap before involving their people beyond a pilot effort.

To focus our TPM program, Agilent created the following vision, strategy, and tactical plan.

AGILENT'S MANUFACTURING VISION

Our vision is what we see ourselves becoming in the future. At Agilent, we have a three-part vision for our manufacturing future:

1. Our equipment does not break down.
2. Our equipment runs at world-class levels of productivity.
3. Our people work at improving their own work areas as a routine part of their daily jobs.

AGILENT'S MANUFACTURING STRATEGY

Agilent's strategy for achieving our manufacturing vision consists primarily of:

- Getting our people out of the business of repairing broken machines, and
- Getting our people into the business of maintaining properly operating machines that assure reliable production capacity every single day.

AGILENT'S MANUFACTURING TACTICAL PLAN

We develop the means for maintaining reliable production capacity through the activities of a *small, cross-departmental action team* on each machine set.

- Each team implements the five TPM steps to improve the productivity of its equipment.
- TPM implementation is guided and supported by a steering committee of top managers.

10

Focusing Machine Improvement Activities

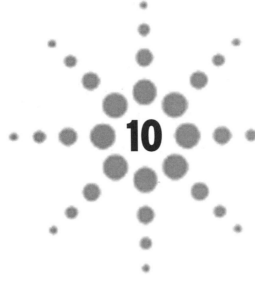

10

Many companies share a common misconception about TPM. They believe they need only apply their TPM activities to a few machines—the bottleneck machines and a few others that have high failure rates. They may also believe that they only need to resolve some of their most severe machine problems in order to achieve world-class productivity levels. This approach to TPM assumes that TPM activities are project methods that can be applied selectively, but this is not how TPM improvement methods work. Implementing TPM effectively means changing the mindset and behavior of everyone in the organization—technicians, operators, engineers, and managers alike. This cannot be done selectively on a few machines. The organization cannot operate under a certain maintenance culture on one set of machines and under another maintenance culture on others.

People in the organization either change their culture with TPM activities or they do not. TPM programs are change programs for people, not just for machines. The only way to change successfully is to change across the board. Either make TPM principles part of the way your people work and the way your company operates its factory, or do not attempt to implement them at all.

However, this does not mean that you cannot focus your TPM activities as you begin implementation. Most people recognize the need for focused improvement efforts for several reasons:

1. Resource limitations prohibit simultaneously attacking every problem on every machine, all at the same time.
2. It takes time to teach people in the organization new methods for making machine improvements. TPM action teams can be created and taught only at a constrained rate.

3. Some machine problems are more harmful to business results than others. So it is entirely reasonable to focus TPM improvement activities first in the areas that pay off the most.

Focusing TPM action teams can be confusing. TPM methods are applied differently than standard improvement methods; both have their uses, but it is important to understand the differences between them.

- Improvement teams can take a Pareto approach to solving machine problems by identifying the most significant losses and using any and all means possible to resolve them.
- Improvement teams can take a TPM approach to solving machine problems by applying the five-step TPM process, largely without regard to the current machine Pareto.

THE PARETO APPROACH

An example from Agilent's experience with the Pareto approach to machine improvement follows.

Data revealed that availability was the largest productivity loss on a certain machine, caused by unexpected breakdowns, as shown in Figure 10-1.

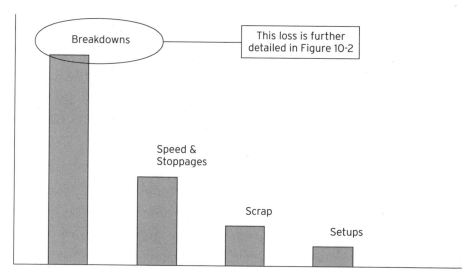

Figure 10-1. Pareto chart of machine productivity loss

When we collected further data to focus on the largest Pareto item, it revealed that most of the breakdowns were caused by wafer-handling failures, as shown in Figure 10-2.

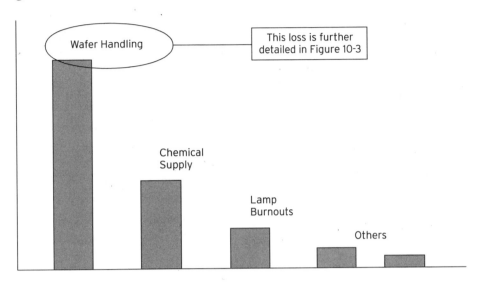

Figure 10-2. Pareto chart of machine breakdowns

By delving deeper into the wafer-handling failure data, another level of Pareto analysis was achieved, which identified the number one cause of the number one failure of the number one machine loss—broken transfer belts, as shown in Figure 10-3.

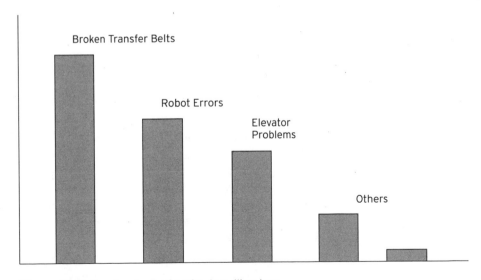

Figure 10-3. Pareto chart of wafer-handling loss

This machine's improvement team began by tackling the largest single failure issue and using the easiest means possible to resolve it. In this particular case, the Agilent team simply chose to halve the PM interval for wafer transfer belt replacements. One result was that belt failures were significantly reduced, so the team appeared to be successful. However, another result was that belt costs, machine downtime for belt replacements, and tech time for changing belts were actually increased by this "solution." Even if this team continues its activities by attacking the next-level problem—and then the next one after that—using this approach, their rate of progress will be very slow. Following this method for years will not result in world-class machine performance.

Ten years of experience has taught Agilent that attacking a specific problem with this type of focused improvement method paid off only when the problem was a large, significant one. The Pareto approach is a good way to begin tackling a handful of very significant problems on the factory floor, but once these large problems have been reduced, teams with a Pareto focus become much less effective. If this type of focus team is all you ever form, machines will create new failure mechanisms faster than your focus teams can eliminate them. Also, maintenance costs are likely to continually increase.

 The Pareto approach to eliminating equipment failures is a good way to approach a small number of significant issues on the factory floor. However, because it does not attack the root cause of equipment loss, it is not a method that can, by itself, help a company achieve world-class levels of equipment productivity.

THE TPM APPROACH

Implementing the TPM steps on a machine is a better method for long-term improvement than continually attacking Pareto issues one at a time. Agilent found that the TPM approach increased machine productivity much more significantly than the Pareto approach. TPM activities also provided machine problem solutions that lowered maintenance costs instead of increasing them. This is partially because TPM methods provide a larger perspective of the factory floor operation:

- *There are no unimportant machines on the factory floor.* Even machines with excess production capacity are bottleneck feeders. If they go down, flow of material into critical bottleneck machines ceases. Reliability problems with any machine on the factory floor have negative business consequences.

- *There are no unimportant parts in a machine or unimportant activities performed by humans on a machine.* The failure of any one part or adjustment to perform its function reliably can make the entire machine behave poorly.
- *Productivity losses in all equipment are caused by many of the same root causes.* A Pareto approach to equipment improvement only attacks one failure at a time, without any global perspective. TPM methods compound rates of productivity improvement by attacking root causes that are common to most equipment failures.

Focusing TPM Steps

It is possible for a TPM team to systematically apply TPM steps to a machine and still focus on the worst machine problems first. TPM methods can be applied to machine subassemblies in very much the same way they can be applied to an entire machine. To accomplish this, a team simply needs to prioritize the order in which they apply TPM steps to the machine's subassemblies, the most troublesome areas of the machine receiving TPM step activities first. Also, working with a small number of subsystems at a time, instead of the entire machine, breaks down the TPM implementation into smaller, more manageable units for TPM action teams. The team must remember, however, that their ultimate goal is to apply TPM step activities to *every* area of the machine.

For example, uniformity problems in an oxide deposition machine could be attacked by applying the five TPM steps to the subassemblies most associated with deposition uniformity, such as the deposition chamber and gas distribution systems. TPM activities on subsystems less likely to be related to uniformity problems could be postponed; these might include pneumatic and wafer-handling components and other subassemblies not directly involved in the deposition process.

To take this approach, identify all subassemblies in a machine such that every single piece of machine hardware is included in one of the subassemblies defined—usually no more than a dozen subsystems per machine. Next, identify the most difficult problems being experienced on the machine and the subassemblies most associated with these problems. Create a subassembly table like the one in Table 10-1. List the subassemblies by priority, with the most troublesome first and the least troublesome at the bottom of the list. An Agilent machine is used in the following example. CVD chambers are our most troublesome area in this machine, control systems our least troublesome.

Table 10-1

Subassembly Priority	Step 1	Step 2	Step 3	Step 4	Step 5
CVD chambers					
Wafer-handling systems					
Gas systems					
Load lock chamber					
Vacuum systems					
Etch chambers					
Pneumatic systems					
HX systems					
RF systems					
Control systems					

Since the CVD chambers and the wafer-handling system are causing the most problems with this machine, we might choose to apply the five TPM implementation steps to the most troublesome areas of this machine first. Table 10-2 indicates this approach, applying Steps 1 through 5 to these two machine subassemblies. (An **X** indicates completion of a TPM step.)

Table 10-2

Subassembly Priority	Step 1	Step 2	Step 3	Step 4	Step 5
CVD chambers	X	X	X	X	X
Wafer-handling systems	X	X	X	X	X
Gas systems					
Load lock chamber					
Vacuum systems					
Etch chambers					
Pneumatic systems					
HX systems					
RF systems					
Control systems					

Once the problems with the most troublesome subassemblies have been greatly reduced, we continue to apply TPM implementation steps to the machine in the same pattern—applying all TPM steps to a small group of machine subassemblies, one at a time in accordance with their established priority. Since the gas system and the load lock chamber are the next most troublesome areas of the machine, we can next apply the five TPM steps to them (Table 10-3).

Table 10-3

Subassembly Priority	Step 1	Step 2	Step 3	Step 4	Step 5
CVD chambers	X	X	X	X	X
Wafer-handling systems	X	X	X	X	X
Gas systems	**X**	**X**	**X**	**X**	**X**
Load lock chamber	**X**	**X**	**X**	**X**	**X**
Vacuum systems					
Etch chambers					
Pneumatic systems					
HX systems					
RF systems					
Control systems					

Once we have eliminated the machine's worst problems, we can simply apply TPM steps to the rest of the machine. In Table 10-4, Steps 1 and 2 are next applied to the remaining subassemblies.

Table 10-4

Subassembly Priority	Step 1	Step 2	Step 3	Step 4	Step 5
CVD chambers	X	X	X	X	X
Wafer-handling systems	X	X	X	X	X
Gas systems	X	X	X	X	X
Load lock chamber	X	X	X	X	X
Vacuum systems	**X**	**X**			
Etch chambers	**X**	**X**			
Pneumatic systems	**X**	**X**			
HX systems	**X**	**X**			
RF systems	**X**	**X**			
Control systems	**X**	**X**			

Remember, the ultimate goal is to apply all five TPM steps to every subassembly in the machine, using the order of prioritization that makes the most sense. A completed TPM activity on our example machine would have all the following activities completed (Table 10-5).

Table 10-5

Subassembly Priority	Step 1	Step 2	Step 3	Step 4	Step 5
CVD chambers	X	X	X	X	X
Wafer-handling systems	X	X	X	X	X
Gas systems	X	X	X	X	X
Load lock chamber	X	X	X	X	X
Vacuum systems	X	X	X	X	X
Etch chambers	X	X	X	X	X
Pneumatic systems	X	X	X	X	X
HX systems	X	X	X	X	X
RF systems	X	X	X	X	X
Control systems	X	X	X	X	X

It is a good idea to display this matrix on a team activity board to track the implementation progress of TPM activities on the machines.

11

Improvement Metrics

11

All TPM activities must be aimed at improving the most significant manufacturing metrics; improvements in trivial metrics will not systematically improve business results. The metrics to be improved must be very carefully specified from a system point of view, and a measurement plan must be put into place. Action teams must be held accountable for improving the metrics that their managers have identified as the most important.

As mentioned in Part I, every manufacturing company strives to improve its TQRDC customer satisfaction. But on the manufacturing floor, machine metrics that factory workers can relate to are more important. Business metrics needs to be broken down into factory floor and machine performance metrics. Once defined, these metrics should be measured and assigned goals for the future.

Metrics commonly used in many manufacturing operations include the following:

- Machine MTTF (mean time to failure)
- Machine MTTR (mean time to repair)
- Product defect rate
- Number of machine failures
- Machine availability
- Number of machine minor stoppages
- Product "move" rates (the pace at which product moves through various phases of the manufacturing process)
- Production rates (good products produced per hour, shift, day, or month)
- Machine setup times

- Product cycle time (the time taken for a product to go from the first step in the production through the last one)
- The ratio of maintenance time spent on preventive maintenance to that spent on breakdown maintenance
- Cp$_k$ trend (a measure of process quality control—Goal > 1.33)
- Number of improvements completed (for example, countermeasures implemented against recurring machine failures)

Performance metrics are useful to action teams in two ways:

- They provide focus for a team, guiding them toward working on the biggest productivity losses first.
- They provide a report card, letting the team and its managers know how fast productivity gains are being generated. This feedback is critical. Successful teams need to know that they are being successful. Teams that are not achieving results need to examine their activities to discover why. In Agilent's experience, most teams that are struggling are not implementing TPM methods accurately. They either ignore TPM methods, attacking problems with their own approaches, or carry out TPM methods without paying much attention to detail.

Agilent's T pilot has experimented with many different factory performance metrics. The following describe some that we have found to be most useful.

OVERALL EQUIPMENT EFFECTIVENESS (OEE)

The so-called "six big productivity losses" have already been described.[1] Any one of these six losses can be used individually as a performance metric, as described in Step 5, or they can be combined into a single metric called overall equipment effectiveness (OEE). These six losses—described for a process industry such as an IC fab—are:

- Availability losses
  - Machine breakdowns
  - Machine setup and adjustment times
- Quality losses
  - Product scrap
  - Low product yields
- Performance losses
  - Minor machine stoppages
  - Reduced machine speed

[1]For more information, see "Productivity Analysis: The Six Big Productivity Losses" on page 228.

Losses can be expressed in either of two ways—what you lose or what you have remaining. Machine availability might be expressed as a loss ("10 percent down") or by a gain ("90 percent up"). To calculate overall equipment effectiveness, we will use the latter, and refer to these values as "rates": availability rates, performance (speed) rates, and quality rates. For example, if 1 percent of our product contains quality defects, the quality rate is 99 percent, or 0.99.

Overall equipment effectiveness is simply the product of the six big losses, and it is most commonly expressed by the following formula:

OEE = availability rate × performance rate × quality rate

- The availability rate is the fraction of the scheduled operating time that excludes breakdown and setup losses.
  - The performance rate is the ratio of the actual production rate of the machine to the theoretical production rate, accounting for both minor machine stoppages and speed losses.
  - The quality rate is the ratio of good product produced to the total number of product produced, accounting for scrap and yield losses.

The OEE concept is illustrated in Figure 11-1.

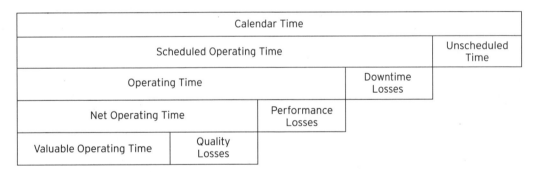

Figure 11-1. A visual representation of the OEE concept

OEE is essentially "Valuable Operating Time" divided by either "Scheduled Operating Time" or "Calendar Time." Each organization must decide for itself which divisor to use for the OEE metric. Agilent uses "Scheduled Operating Time" as the basis for OEE calculations. This choice excludes from our OEE measure the excess capacity in non-bottleneck machines. It allows us to focus on the actual performance losses of these machines, without including the fact that we own nine machines when we only need 8-1/2 of them to meet our capacity needs. However, "Calendar Time" can be used as the basis of the OEE metric if it is desirable to have excess machine capacity appear as a loss in the productivity metric.

Another choice in deciding how to produce an OEE metric is whether to categorize scheduled maintenance time as a downtime loss or as unscheduled time. Again, this depends on the productivity losses one wishes the OEE value to represent. Agilent has chosen to include scheduled maintenance time as a downtime loss.

Typical OEE values found in an average manufacturing plant are 35–45 percent. Plants that have made good productivity gains may operate with OEE values of 50–70 percent. World-class manufacturing plants operate with OEE values of about 85 percent.

TOTAL PRODUCTION RATIO

In some manufacturing situations, it is reasonable to implement the OEE metric. However, OEE is often very difficult to measure, even on simple integrated manufacturing lines. Fortunately, there is a more easily obtainable metric, the total production ratio (TPR), which is closely related to OEE. TPR tracks OEE consistently because it is derived by multiplying the definition for OEE by a term called the operating rate, and the operating rate is—for all practical purposes—nearly a constant. Figure 11-2 shows OEE and TPR data for a manufacturing line over a period of two years. TPR is consistently less than OEE because overtime was used regularly during this period, making the operating rate value less than 1.0.

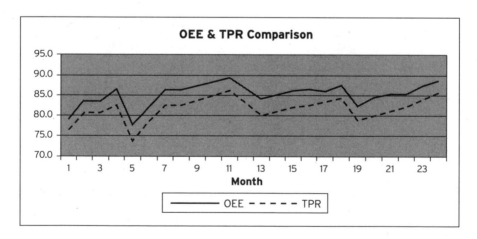

Figure 11-2. TPR tracks OEE very consistently

A simple derivation of TPR follows, including definitions of terms used in the derivation. It is not necessary to understand all the terms or to follow the derivation in exact detail, since at no other point in this TPM instruction are most of these terms ever applied. This information is only offered as additional evidence that TPR is a valid substitute for OEE because they are mathematically linked.

TPR Derivation Terms

OEE = Availability rate x performance rate × quality rate.

Availability rate is the fraction of scheduled operating time that excludes breakdown and setup losses.

Performance rate is the ratio of the basic product time to the actual average product time, accounting for both minor machine stoppages and speed losses. (Although this definition of performance rate is worded differently than the one offered in the previous section on OEE, the definitions are mathematically identical.)

Basic product time is the theoretical time between parts coming off the end of a machine or line. (Product time might be minutes per part, and is the reciprocal of the rate at which parts are produced by the machine—for example, parts per minute.)

Actual product time is the actual average time between parts coming off the end of a machine or line.

Quality rate is the ratio of good product produced to the total quantity of product produced, accounting for scrap and yield losses.

Operating rate is defined as the scheduled time divided by the working time.

Scheduled time is the total time scheduled for production operation during a work period, less planned or necessary downtime, such as breaks.

Working time is the scheduled time, plus any overtime that the line may have actually been run.

Operation time is the scheduled time less downtime.

Downtime is the time a machine is unavailable for planned production because of breakdowns and other unplanned stoppages.

Net operation time is the time the equipment operated at nominal speed, accounting for minor stoppages or other speed losses.

Valuable operation time is the portion of net operation time during which acceptable quality products are manufactured.

Total production is the count of all product manufactured in a given production period.

Total quality production is the count of all good product manufactured in a given production period.

Given these definitions, the derivation of TPR follows:

$$\text{TPR} = (\text{Operating Rate}) \times (\text{OEE})$$

Expanding the definition of OEE yields:

$$\text{TPR} = (\text{Operating Rate}) \times (\text{Available Rate}) \times (\text{Performance Rate}) \times (\text{Quality Rate}) \times 100\% \tag{1}$$

More detailed equations for these terms include the following:

$$\text{Operating Rate} = \frac{\text{Scheduled Time}}{\text{Working Time}} \tag{2}$$

$$\text{Availability Rate} = \frac{(\text{Scheduled Time} - \text{Downtime})}{(\text{Scheduled Time})}$$

$$\text{Therefore, Availability Rate} = \frac{(\text{Operation Time})}{(\text{Scheduled Time})} \tag{3}$$

$$\text{Performance Rate} = \frac{(\text{Basic Product Time} \times \text{Total Production})}{(\text{Operation Time})}$$

$$\text{Therefore, Performance Rate} = \frac{(\text{Net Operation Time})}{(\text{Operation Time})} \tag{4}$$

$$\text{Quality Rate} = \frac{(\text{Total Quality Production})}{(\text{Total Production})}$$

$$\text{Quality Rate} = \frac{(\text{Total Quality Production} \times \text{Basic Product Time})}{(\text{Total Production} \times \text{Basic Product Time})}$$

$$\text{Therefore, Quality Rate} = \frac{(\text{Valuable Operation Time})}{(\text{Net Operation Time})} \tag{5}$$

Substituting equations 2, 3, 4, and 5 into equation 1 yields:

$$TPR = \frac{\text{(Scheduled Time)}}{\text{(Working Time)}} \times \frac{\text{(Operation Time)}}{\text{(Scheduled Time)}} \times \frac{\text{(Net Operation Time)}}{\text{(Operation Time)}} \times \frac{\text{(Valuable Operation Time)}}{\text{(Net Operation Time)}} \times 100\%$$

Canceling scheduled time, operation time, and net operation time from the numerator and denominator leaves:

$$TPR = \frac{\text{(Valuable Operation Time)}}{\text{(Working Time)}} \times 100\%$$

Redefining the numerator yields the useful equation defining TPR:

$$TPR = \frac{\text{(Total Quality Production} \times \text{Basic Product Time)}}{\text{(Working Time)}} \times 100\%$$

This very simple equation for total production ratio can often be used instead of the more complex OEE. Of course, this metric is just a report card indicating the results of improvement efforts. It will not tell TPM action teams which losses must be reduced to improve equipment productivity.

As an example of TPR, consider a manufacturing line scheduled to operate from 8:00 A.M. to 5:00 P.M., including a one-hour lunch break. During the remaining 8 hours, the line operators take two 15-minute breaks. Therefore, the actual scheduled production time for the line is 7.5 hours. When operating properly, the line can produce a part every 1.27 minutes, the basic product time for the line.

During one routine day, the line broke down twice, had seven minor stoppages, and produced some poor-quality parts that had to be scrapped. No scheduled maintenance was performed. Since no overtime was used to make up for the losses, the working time of the line for this period is the same as the scheduled time—7.5 hours, or 450 minutes. Despite the line's problems, it still produced 276 good products during this shift. The TPR for this shift is therefore:

$$TPR = \frac{(276 \times 1.27)}{(450)} \times 100\% = 78\%$$

Should this line actually produce a part every 1.27 minutes as it is expected to, and run perfectly for the entire 450-minute period, it would produce 354 parts during this time (450/1.27). In this case, the total production ratio would be, of course, 100 percent:

$$TPR = \frac{(354 \times 1.27)}{(450)} \times 100\% = 100\%$$

Any operator can make this calculation and record it daily on the manufacturing line's TPR performance chart. If the TPM action team is carrying out TPM activities properly, this metric should continually improve.

NUMBER OF MACHINE FAILURES

Measuring the number of machine failures is a relatively simple metric that can be implemented easily. However, it has some weaknesses. Some failures last only a few minutes; others may last for days. A failure of a bottleneck machine may have more impact on business metrics than a similar failure of a non-bottleneck machine. Failures of equal duration on non-bottleneck machines affect material flow to bottleneck equipment differently. A one-hour shutdown on one machine may starve the bottleneck, while a one-hour shutdown on another machine may have little effect on the flow of material into the bottleneck. Still, this is a good place to start measuring productivity losses if data-collection capabilities are limited. Machine operators, using tick marks on a clipboard or chart at the machine, can track the number of failures. As an action team works to reduce the number of equipment failures, many other productivity metrics—for example, machine availability—will likely get improved as well, whether or not they are actually measured.

DURATION OF MACHINE FAILURES

Another useful performance metric is the duration of each machine failure—not the average or mean, such as MTTR, but data on the actual time that the machine was out of service. The longer a machine is out of service, the more it disrupts productive line flow. A sample of downtime durations over the course of a month for a machine is shown in Figure 11-3. This measure visualizes the number of failures per month and also provides insight into how disruptive each down event was.

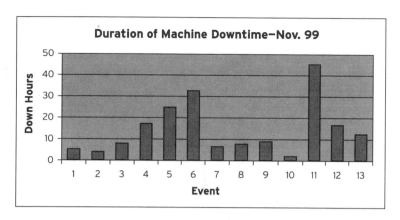

Figure 11-3. A monthly record of the downtime duration for each failure event during the month

MACHINE FAILURE HISTOGRAM

Another simple way to visualize machine failure is to make a histogram of the number and duration of each failure. Figure 11-4 shows data from a machine that has unscheduled downtime for a few hours about every three days. But three times during this month, the machine failed and remained out of service for considerably longer—once, for almost 30 hours.

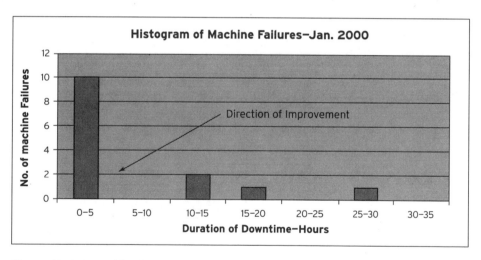

Figure 11-4. A machine failure histogram indicating number of failures and failure duration

Machine improvement on this type of histogram chart is reflected by the bars moving toward the lower-left corner; the height of the bars becomes shorter and the bars near the right side of the chart will move to the left as the downtime duration decreases. Ultimately, the x- and y-axis scales become smaller and smaller numbers.

MACHINE CAPACITY ASSURANCE

Measurements of machine capacity provide a useful metric for understanding the performance of both bottleneck and non-bottleneck machines. This metric is simply the number of products actually produced by a machine during a day divided by one minus the fraction of unscheduled machine time during that day. (The time period need not be a day. It can be any time period. The shorter the time period, the more visible are fluctuations in machine reliability.)

$$\text{Daily Machine Capacity} = \frac{\text{Actual Daily Production}}{(1 - \text{Unscheduled Machine Time})}$$

Unscheduled machine time in this equation is defined as the fraction of the time period that the machine was available, or "up," but was idle and had no WIP in front of it. For machines where WIP is always available for production, there is no unscheduled time, so the current machine capacity is, of course, the actual number of products produced divided by one. But suppose a machine is required only for half the time to keep up with the bottleneck. Assuming that it is always up, it will be unscheduled half the time. Therefore, its daily machine capacity will be twice its actual daily production.

Non-bottleneck machines have excess capacity and therefore will be idle sometimes, since the flow of material through them is limited by the bottleneck. Their job is to keep material feeding the bottleneck machines reliably. The degree to which these machines are unreliable usually dictates how badly they hurt factory throughput. Measuring their daily capacity and analyzing the variation is one way to visualize the machines' reliability. If a machine works well for three months and then suddenly breaks down for a week, the first problem to attack is the cause of that breakdown.

 As long as the capacity of every group of non-bottleneck machines is kept above the bottleneck's supply needs, the bottleneck will never starve for material.

The period of time over which reliable machine capacity assurance is required depends on the nature of the manufacturing line and the current state of equipment reliability. The goal is to improve equipment reliability to meet scheduled capacity over a shorter and shorter period of time.

Agilent is currently being challenged to provide reliable capacity assurance on a daily level. Some factories with much shorter product cycle times might be challenging themselves to produce these same results on a shift or hourly level. One can imagine that automobile assembly plants might be working toward reliable capacity assurance at every assembly station on a minute-by-minute level.

No matter what the current challenge, the ultimate goal of equipment reliability is that machines never break down. The more reliably equipment approaches this goal, the less excess WIP will be required to ensure factory throughput. As equipment reliability is improved, excess WIP levels can be reduced, improving factory productivity, and lowering factory cycle time and product cost.

Figure 11-5 shows the average capacity for a machine compared to its scheduled needs over a three-year period. Averaging daily capacity over a year provides clear evidence that this is a non-bottleneck machine, but also falsely implies that the machine is not a factory productivity limiter. Data on the reliability of the machine on a shorter time basis are required. In this three-year period, the daily requirement is 105 parts. During this time, the machine was capable, on average, of producing 123 parts per day.

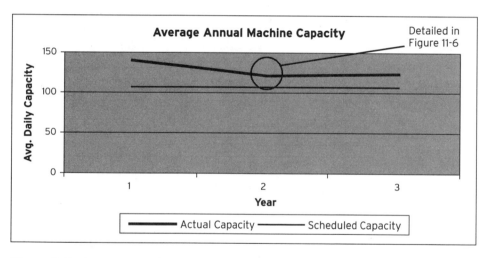

Figure 11-5. Average machine capacity plotted on an annual basis

Figure 11-6 is a "blow up" of year 2 from Figure 11-5—the year that appeared to have the lowest machine capacity. It shows daily machine capacity averaged over each month of the year. A weakness that is not apparent in the annual averages is starting to appear in the monthly averages. Although the machine averaged a capacity of 123 parts per day over the course of the year, during the fifth month it was barely able to meet the scheduled capacity needs, averaging a capability of only 107 parts per day—barely more than the required production schedule.

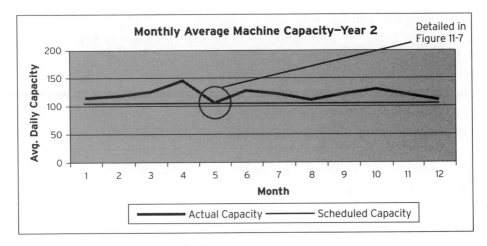

Figure 11-6. Machine capacity plotted on a monthly basis

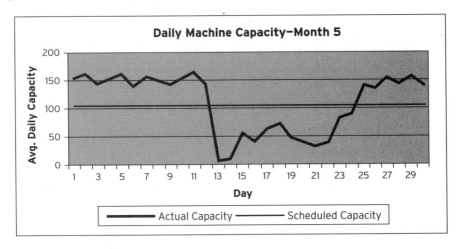

Figure 11-7. Machine capacity plotted on a daily basis

If we expand the data from month five and show the daily machine capacity every single day of that month, a weakness in the machine's ability to provide production capacity is finally revealed (see Figure 11-7). Although the average machine capacity for the month exceeded the scheduled capacity—107 parts per day versus the 105 required—the machine fell short of meeting its scheduled capacity needs for 10 days in a row during the month. This type of machine performance is very disruptive to a manufacturing line and causes lost productivity.

Machine capacity data should be plotted over a length of time that is significant for production needs. Long-term averaging is of little use, because it masks reliability problems that have significant negative impacts on manufacturing productivity.

Figures 11-8 through 11-11 show different daily capacity charts for various machines. The patterns on these charts provide insights into the machines' capacity and reliability. The variable line plots actual daily machine capacity. The straight line indicates the machine's requirement to produce 105 parts per day to meet the factory production schedule. An interpretation is provided for each pattern.

The capacity chart in Figure 11-8 shows a machine that behaves (at first glance) like a fairly reliable bottleneck machine. Its average capacity of 105 parts per day barely meets the plant's needs for the machine. However, some days it produces more than its quota, some days less, indicating that reliability improvements would increase this bottleneck machine's capacity. The machine produced 120 parts on at least one day of this month, indicating a possible productivity gain of 14 percent just from reliability improvements. All the productivity losses on this machine should be attacked, because productivity gains on this machine will improve utilization of all tools in the factory.

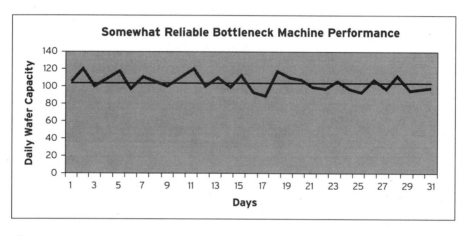

Figure 11-8. Daily wafer capacity chart of a bottleneck machine

Figure 11-9 shows a fairly reliable non-bottleneck machine set consisting of two machines. This machine set has more capacity than is required to meet factory scheduling needs, and it is reliable enough to meet these needs every single day of the month. However, the excess capacity of these machines is a productivity loss in itself. Perhaps, instead of owning two of these machines operating at current productivity levels, increased productivity would allow the shop to produce the same number of parts with only one machine. This type of productivity improvement in non-bottleneck machines is another way of reducing factory operating costs.

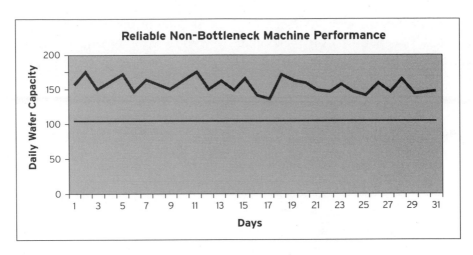

Figure 11-9. Daily wafer capacity chart of a non-bottleneck machine

Figure 11-10 shows the daily capacity performance of a non-bottleneck machine with some reliability problems. The machine has enough excess capacity to meet factory schedule requirements, but a few days each month, it has difficulties that require it to short-ship material. Having some excess WIP between this machine and a downstream bottleneck machine usually compensates for this type of problem. However, such excess WIP is costly for many reasons. The problems causing this machine to underperform should be eliminated.

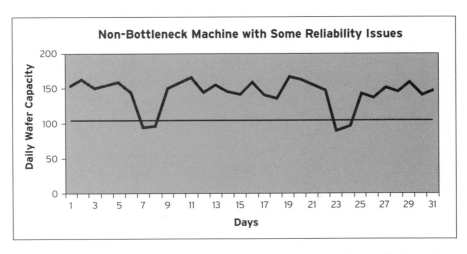

Figure 11-10. Daily wafer capacity chart of a somewhat unreliable non-bottleneck machine

Figure 11-11 shows a non-bottleneck machine with a severe reliability issue. The machine has excess capacity to meet factory scheduling needs but had at least one failure mechanism that greatly reduced its production capacity for nearly 10 days in a row. This is one of the worst types of machines on a factory floor. The disruption in the flow of material causes tremendous factory losses. Resolving this problem must be a very high priority.

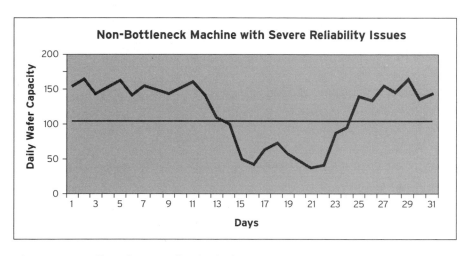

Figure 11-11. Daily wafer capacity chart of a very unreliable non-bottleneck machine

LINEAR PRODUCT FLOW

Another metric used to objectively measure the performance of machines on a manufacturing floor is one that measures the rate at which material flows into a key machine on the line. To understand rate, imagine a perfectly reliable set of factory machines making the same product over and over again, producing a perfectly linear flow of material through the line. In the real world, this rarely happens for long. Deviations in this linear flow of material are caused by reliability problems in equipment throughout the line. Measurements of non-linear flows of material at a strategic point in the line can be used to identify problems with machines that cause the most severe losses in factory linearity and productivity.

This principle is illustrated in a simple factory model made of factory floor building blocks. Each building block might be a single machine or an arrangement of machines as shown in Figure 11-12.

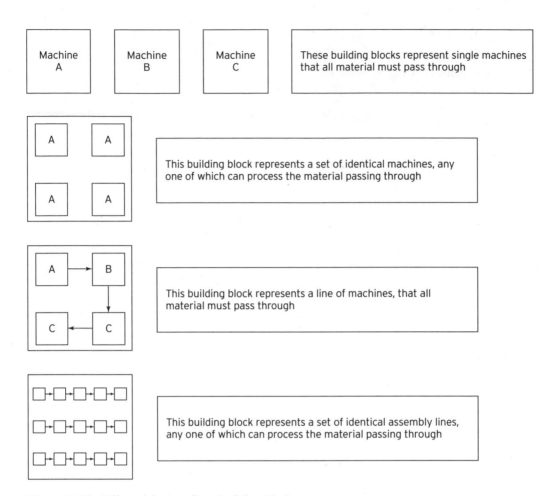

Figure 11-12. Different factory floor building blocks

A factory can now be illustrated as a sequence of manufacturing operations, each one represented by one of the above building blocks. The bottleneck operation is the one with the lowest design throughput capacity—the third operation in this example.

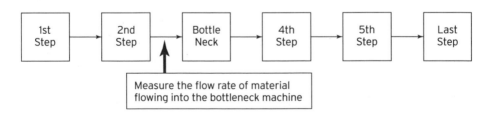

When the bottleneck machine is identified, the rate at which material flows into it can be measured. (If the bottleneck machine is unknown, pick a strategic location—either the suspected bottleneck or somewhere in the middle of the line where measurements can be taken reasonably well.)

In a perfect manufacturing line, the rate of material at the measurement point would be linear and of sufficient quantity to meet scheduled output. However, the actual flow rate will be somewhat non-linear depending on problems that occur on the line.

Measurements of incoming material were made at the bottleneck machine (see Table 11-1). (Operators in the area placed a tick mark in the second column of the chart—"Incoming Lot Count"—every time they received a lot of incoming material.)

Table 11-1

Time Period	Incoming Lot Count (Normal = 10)	Hourly Total	Shortage	Source of Shortage
8:00-9:00	IIIIIIIII	9	1	1st Step
9:00-10:00	IIIII	5	5	1st Step
10:00-11:00	IIIIIIIIIIIIII	14		
11:00-12:00	IIIIIIIII	9	1	5th Step
12:00-1:00	IIIIIIIIII	10		
1:00-2:00	IIIIIIIIIIII	12		
2:00-3:00	IIIIIIIIII	10		
3:00-4:00	IIIIIIIIIIIII	13		
	Total	82		
	Average Hourly Rate	10-1/4		

As can be seen by the data summary at the bottom of the third column, the average quantity of material actually shipped into the bottleneck machine was slightly greater than the material needs for the production day. That is, the bottleneck machine actually received an average rate of material that day of 10-1/4 lots per hour when scheduled to receive only 10 lots per hour. This might make one think that this line was actually running better than planned. However, the flow of material was somewhat non-linear, being short during the early part of the shift and "catching up" during the later hours. The bottleneck machine could have been kept running during the short-delivery periods, of course, by keeping a WIP buffer of sufficient size in front of it to fill in for the short-shipped materials.

But the goal of this technique is to identify and resolve problems to improve the productivity of the line. Therefore, every hour that a shortfall occurs, an operator will "take a walk" to discover the cause of the shortage. If the line is a "pull-type" JIT line, and the incoming shortage was the result of a lack of pull cards because of material backing up downstream from the bottleneck machine, then the operator will take a walk downstream, going from area to area in the process flow to pinpoint the slowdown of material. If the bottleneck machine is "pulling" material that is not being delivered, then the operator will take a walk upstream.

In a "push-type" line, the operator will always walk upstream to find the source of the short material deliveries to the bottleneck machine. The operator will note the area causing the problem in the last column of the data chart.

In this particular example, a "pull-type line," the primary shortages in the morning were caused by problems in the first manufacturing step. Later in the day, a small shortage was caused by the fifth manufacturing operation.

Two actions can result from these measurements. First, immediate action can be taken to focus resources on areas that are disturbing the steady flow of material into the bottleneck machine. Second, the problem areas that most interrupt material flow on the factory floor can be identified.

Table 11-2 summarizes the number of times a machine area was identified as a problem in supplying linear material to the bottleneck machine over a period of three months.

Table 11-2

Operation	Number of Identified Problems
1st Step	127
2nd Step	14
Bottleneck	9
4th Step	34
5th Step	67
Last Step	2

Table 11-2 clearly indicates that, in this case, the bottleneck machine itself rarely limits material flow. Rather, the first step in the production process most often is the limiter, even though this machine has more capacity than the bottleneck. This indicates that the first machine has serious reliability problems. The data also indicate that fourth- and fifth-step machines have reliability problems.

These data allow us to focus limited resources on the most constraining areas, which are easily identified by this rather simple method.

 The "secret" to the "linear product flow" method is to "take a walk" and "go look" for the cause of the line disruption in real time—*when the problem is actually occurring and can be easily identified.*

Just as in the example above, the bottleneck machine in Agilent's IC fab is not the machine that most often limits factory output. The bottleneck machine is often starved for material, and the time it spends idle because of lack of work can never be recovered. Even though all other machines have more capacity than the bottleneck machine, their inability to reliably deliver material to the bottleneck machine in a reasonably linear fashion is the true root cause of most lost factory output.

FACTORY OUTPUT AND PRODUCT COST

Factory productivity is—in its simplest form—factory output divided by factory cost. Therefore, increased factory throughput that is sold to customers is, perhaps, the most telling measure of the results of TPM activities, despite other metrics that might show improvement. Always remember to include total production output and product cost as key business and TPM metrics.

Without an eye on system-level results, teams tend to produce local optimizations instead of system optimizations. Often, local optimizations will not improve factory output at all. For example, if the speed of one very reliable non-bottleneck machine were increased somewhat, but not so much that the number of this type of machine could be reduced, neither factory output nor business results would benefit; on the contrary, factory resources and people's time would be consumed without delivery of any business improvement. TPM teams engaged in improvement activities must account for the resources that they expend.

12

Managing New Behaviors

As stated earlier, TPM activities require behavior changes. To manage TPM teams effectively, managers must learn the rudiments of managing behavior changes, also known as performance management. The following section presents some basics on managing new behaviors.[1]

THE ABCS OF BEHAVIOR

People's behavior can be explained at the simplest level by the "ABCs of behavior"—antecedent, behavior, and consequence.

- *Antecedent*: An event preceding a behavior; a cue for the behavior to take place. For instance, a scheduled PM that has become due is the antecedent for a certain tech behavior—performing the PM.
- *Behavior*: The actual performance that a person carries out. Behavior is human activity that is observable to others. It is not a thought or an attitude. For instance, in this example the technician's completion of the PM is the expected behavior.
- *Consequence*: What happens to a person during or after the behavior as a result of performing it. For instance, completing the PM might result in several consequences to the technician who performed the work. The technician may perceive these consequences as either positive or negative.

[1]For more information on this subject, refer to *Performance Management* by Aubrey C. Daniels, Performance Management Publications, Tucker, Georgia, ISBN 0-937100-01-3.

It might seem that the antecedent causes certain behaviors to be repeated, but in fact it is the consequences following the behavior that cause the behavior to be carried out in the future when asked for. The antecedent might only get the behavior to occur once or a very few times when first used. The consequences that are delivered during or after the new behavior determine whether that behavior will be repeated when required. Punishing consequences cause a person to attempt to ignore future antecedents for the behavior. Positive consequences cause the person to respond favorably to future antecedents for the behavior.

Behavior Antecedents

Antecedents in the workplace usually include the 4 Ps to prepare people to carry out any behavior:

- Teach people the *purpose* of the behavior. (For TPM, purpose is defined by the company's vision.)
- Prepare *policies* that provide guidelines for the behavior. (For TPM, policies are defined by the management steering committee.)
- Prepare detailed *procedures* for the behavior. (For TPM, procedures are defined in the factory operating systems, TPM step checklists, and documentation created to support TPM activities.)
- Provide *proficiency* in the behavior through training and practice. (For TPM, proficiency is initiated by our training and skill-development systems, but is developed for the most part by people practicing the behavior.)

Behavior Consequences

There are four types of consequences for behavior, often referred to as follows:

- *R+ (R plus)*: A consequence delivered to the performer that is something the performer desires. For example, a person that you admire and respect telling you, "You did a great job" after you worked very hard on a task.
- *R– (R minus)*: A consequence delivered to the performer that is in the form of a threat. For example, "You have failed to complete the report that I assigned to you. If you don't get that report on my desk by five o'clock today, you will have to stay late to complete it." (Sometimes R- is required to produce a behavior when the normal antecedent fails.)
- *P+ (P plus)*: A punishing consequence, whereby performers receive something they do not want. For example, a person that you admire and respect telling you, "You did a very poor job on that project. I am disappointed in you."

- *P– (P minus)*: An extinction consequence. In other words, the absence of any active consequence—the performer is simply ignored. A mother walking out of a room, leaving a two-year-old child throwing a tantrum on the floor, is extinguishing this behavior by ignoring it. She is offering neither positive reinforcement nor punishment.

The most powerful reward in the workplace is management's positive attention to people doing the work. Ninety percent of a good reinforcement program is positive recognition from the staff's chain of managers and from their peers. The use of money or gifts as rewards for new behaviors should be kept to an absolute minimum. Also, the work itself can be inherently rewarding. For example, accomplishing a PM with less effort that produces better results on the machine is inherently rewarding for most maintenance technicians.

Remember, though, that a consequence is either positive or negative depending on the *perspective of the performer* and not on the perspective of the person offering the consequence. I might recognize someone for an achievement by asking that person to stand up in front of our entire organization during an assembly, intending this as a reward for his or her performance. However, if the person is shy and fearful of public attention, my "reward" may actually be perceived as a punishment. People have different ideas about rewards and punishments, so managers would do well to understand what each team member likes and dislikes before delivering consequences.

SSIP Rules

Consequences should always be delivered by the SSIP rules:

- *Specific*: Tell people exactly why they are being rewarded. For instance, "You did a good job of finding the root cause of our water pump failure on the etchers," rather than "You're doing good work. Keep it up."
- *Sincere*: Be honest about the praise you offer. People have an incredible ability to sense false sincerity. If you don't mean it, don't say it. A reward that a person receives but believes is insincere is probably punishing to them, not rewarding.
- *Immediate*: Offer rewards to a performer as soon as possible after you get the behavior you want—or better yet, during the actual performance of the behavior. Timing is critical. Rewards offered after any significant time has passed after the behavior mean little to the performer.
- *Personal*: Offer personal rewards—from me to you. For example, I might say to you, "I really liked the way you managed to complete your project on time," rather than "The company appreciates your contribution to this project."

People will not engage in new behaviors just because you ask them to. Offering positive reinforcement for engaging in the new behaviors necessary, for TPM requires that managers devote their time to and take a sincere interest in what their people are actually doing.

The leader of one of Agilent's TPM teams made the following table and hung a new copy of it on his office wall nearly every day. This was a very useful tool to remind him to reward people for their contributions to the team's effort.

Team member	What they did today to contribute to the team's progress	The positive consequence that I delivered to them
Mike Verville		
Ed Hockensmith		
Travis Swearingen		
Kevin Funk		
Jerry Akers		
Tim Hailey		

In most cases, the consequences delivered were a sincere "Thank you for a job well done" delivered face to face by the team leader. In some cases, written memos were sent to every manager the person reported to—all the way to the top of the organization—pointing out specifically what good work a person doing.

Having a high-level manager occasionally participate in team activities and observe individual behaviors is another way to deliver consequences to TPM team members to create new behaviors. Significant rewards can be delivered on the spot to help improve the desired behaviors. The following guidelines were used by one of Agilent's TPM team leaders during early TPM team activities.

Behavior Observed	Team Leader Response
Correct behavior based on the instructions given to the team member.	**Reinforce** this behavior by pointing out specifically what the performer is doing well, and let the person know that this is what you want and appreciate.
Good effort at the desired behavior, but improvement is needed.	**Teach** team members how to improve their performance. Let them know exactly what they are doing well, but give guidance for improved performance.
Nonperformance or unwanted behaviors.	**Correct** the improper performance. Tell the team members what they must not do; also, show them what they should be doing and teach them how to do it. Set clear expectations. Deliver punishing consequences if the behavior does not improve, especially if it is abusive to other people or is sabotaging the team's progress.

CELEBRATIONS

Celebrations differ somewhat from positive reinforcement. Reinforcement is delivered according to the SSIP rules, contingent on the correct behavior being performed; celebrations are for everyone to mark a milestone completed by the efforts of the team. The TPM pilot team often went out to lunch to celebrate an accomplishment. We also brought in pizza for the entire shift of people who contributed, so that they could be included in a celebration.

SHAPING BEHAVIOR

Managers should learn another concept, shaping, that helps to develop new behaviors in an organization. In shaping behavior, one starts out rewarding small accomplishments, and then "raises the bar" to higher and higher standards before again rewarding performance. TPM team members should be rewarded for new behaviors, even at a low level of accomplishment; the team will improve its performance as its efforts are being rewarded. Punishing teams for not being perfect will quickly kill new TPM activities—raise the level of expectation only as teams develop higher levels of knowledge and skill.

For example, the first failure analysis done by most Agilent techs was of poor quality. But it was better than what they were doing before, which was repairing broken machines and then doing nothing to prevent the problems from recurring. Thus, even elementary attempts at failure analysis by people doing them for the first time were appropriately rewarded.

PERFORMANCE IMPROVEMENT PROCESS

TPM managers should use a "performance improvement process" to help them promote new behaviors in their organizations. This process consists of the following steps:

1. Pinpoint specifically what is desired. Include:
   - The end results desired
   - The behaviors you believe will produce these results
2. Design and implement an antecedent program to prepare people to carry out their new mission:
   - Tell them what is going to be done
   - Teach them how to do it
3. Design and implement a measurement system. Decide what metrics support the pinpoints. (What must improve?) Include both:
   - Behaviors
   - Results

4. Provide the performers with feedback on the results of the measurements (i.e., let people know how they are doing). This, alone, will often produce improved results.
5. Design a positive-reinforcement program to reward improved behavior. (Deliver consequences that are contingent upon the behavior performed.)
6. Continually review and improve all the previous steps of this performance improvement plan, letting your previous experiences teach you how to accomplish each one in an improved fashion.

13
Team
Activity Boards

13

One of the best ways to provide team accountability and reinforcement, and manage team activities, is to have each team produce an activity board displaying team goals, activities, and measured results. Agilent managers hold monthly audits with the team members in a standing meeting in front of their activity board. This audit is used to provide guidance and positive reinforcement for team accomplishments.

An activity board should consist of three primary sections: Team Charter, Team Activities, and Team Results. The "Team Charter" section should include the names of team members; the machine, machine set, or manufacturing line the team is working on; and the team goal. The goal should state specific improvement metrics and targets. The goal should also include the following charter as well.

The team goal is to establish high-quality, rigorous maintenance and operating procedures using the five steps of TPM to reduce equipment failures and improve manufacturing productivity. The team is currently engaging in Step ___ activities.

The "Team Activities" section of the activity board should include a TPM progress matrix[1] to indicate the current status of the team activities. This section of the activity board should also include a summary of the team's activities and samples of their work, including:

---

[1]An example of a TPM progress table is shown in "Focusing TPM Steps" on page 261.

- Maintenance plans, PM checklists, and maintenance procedures
- 1-Point Lessons
- Failure analyses and implemented countermeasures
- Design and other improvements made by the team

The "Team Results" must include objectively measured trend charts of the specific metrics that the team is designated to improve. In the early stages of a new activity, it may also contain metrics on behaviors such as the number of 1-Point Lessons written or M-tags closed. This section of the activity board may also include stories or lists of team accomplishments.

A typical Agilent activity board is shown in Figure 13-1.

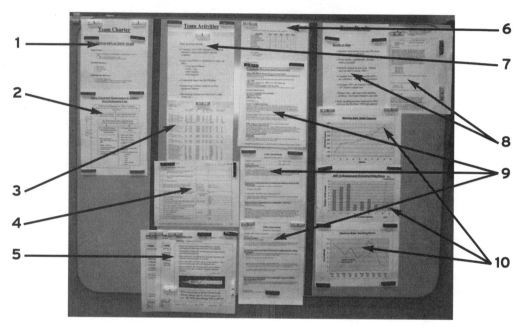

No.	Description
1	Team ID and charter; machine description
2	Five-step TPM process flow chart
3	Master PM plan for the machine
4	Sample PM checklist
5	1-Point Lessons
6	TPM status matrix
7	Description of team activities
8	Description of team results
9	Failure analyses, completed and in progress
10	Machine metric trends

Figure 13-1. A typical Agilent TPM team activity board

14

Sustaining TPM Changes

14

Sometimes, implementing the changes required by TPM can be more difficult than developing the actual TPM activities. Changing people's longstanding behaviors in an organization is always difficult; resistance to change is natural. That is why TPM programs must be driven from top management positions in an organization. TPM is not a grass-roots program that will spread naturally from operator to operator and technician to technician. At Agilent, TPM is managed and driven by a steering committee of top managers. External or internal TPM consultants can help to develop TPM technology, but only managers can deploy TPM activities throughout the organization.

Changes introduced into the organization by TPM activities must be anchored there by becoming an established part of everyone's daily work routine. If TPM is a project—rather than a change process—it will be temporary, and its gains will fade away after the project is completed.

Unfortunately, some PM prize-winning companies have failed to retain all the productivity gains they made with TPM, probably because they did not create an organizational infrastructure to sustain all the new TPM behaviors. Once a TPM "project" is completed, and the TPM office has disbanded, people often stop engaging in the new behaviors that the TPM program established.

This is why an infrastructure support checklist has been included with each TPM implementation step in Part II of this book—to assure that new TPM activities become part of the normal way that people do business in your factory. They must be supported by your normal operating and reward systems.

Successful TPM progress is observable. One tool used by Agilent's TPM Steering Committee to evaluate the evolution of permanent changes taking place in our organization was the "TPM Gap Analysis." This checklist, or a similar one you create, can be used to measure progress toward the specific goals you have set for your TPM program.

TPM GAP ANALYSIS

TPM activities that would be observed in a highly developed TPM organization	Current Score 1-10	Observations/ Improvement Needs
All equipment is maintained "clean and defect free"		
Every machine has a PM plan containing all required PM elements		
Every machine failure is analyzed, and action is taken to prevent recurrence		
Every PM is evaluated for improvements with each PM completion		
All quality components have been identified and are being maintained		
Productivity losses are being measured on every piece of equipment		
Cross-departmental teams make significant advances against equipment productivity losses		
Everyone pursues a well-defined skill development plan		
1-Point Lessons are written and shared "by the thousands"		
Managers lead the way in generating all the above activities		

15
Final Thoughts

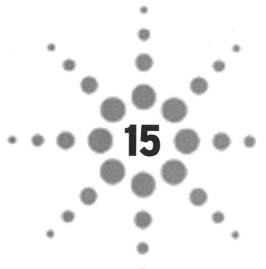

15

The principal goal of any manufacturing business is to produce profit, no matter the product being manufactured. One way to increase profitability is to improve manufacturing productivity, making more product with the same, or a lesser, amount of manufacturing resources. Industries that rely heavily on machines in their manufacturing operations can improve productivity by assuring that their machines can reliably provide scheduled capacity for their needs, hour by hour and day by day. TPM activities are a way of achieving these desired results.

Implementing TPM activities means changing many normal behaviors of all employees—operators, techs, engineers, and managers, alike. This process is difficult, even in a professional environment where people share the same goals. Implementing TPM is hard work, fraught with roadblocks and pitfalls. Management commitment and participation in this change program is imperative if it is going to be successful. If you are not prepared to see through what you have begun, initiating a TPM program will do more harm than good.

Some organizations believe that they must achieve buy-in from their employees before commencing new activities. However, most people will never recognize the merits of TPM activities through hearsay alone. Real buy-in comes after people have experienced positive results from doing something new. When many of Agilent's maintenance techs realized that TPM activities made their job easier—instead of harder, as they had first imagined—they then bought-in to the TPM program and promoted it.

The Agilent TPM steering committee's progress stalled on a number of occasions, each time for a different reason. Some of the problems seemed insurmountable, but persistence—and continuing to take one small step at a time—always saw us through and got us back on track.

We learned several lessons from these difficulties:

- *Be patient!* Don't overload your people. You must proceed at a deliberate pace with the changes your TPM activities bring about.
- *Never give up!* Keep moving forward, no matter what obstacles you may encounter.
- *Stay focused on the road ahead!* Never let anyone or anything deter you from your planned course of action.

Remain Stubborn in Your Pursuits!

Appendices

APPENDIX A

Example of Maintenance Training Material

Maintenance Technician Mini-Course–Fluid Fittings
Used at Agilent's Fort Collins, Colorado, IC Fab

COURSE CONTENTS (ABBREVIATED)

- Pipe thread fittings
- JIC fittings
- SAE fittings
- Fitting adapters
- Swagelok tube fittings

PIPE THREAD FITTINGS

Pipe threads are tapered threads (Figure A-1):

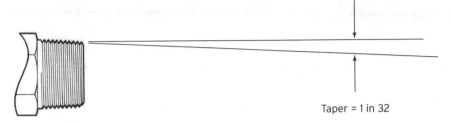

Taper = 1 in 32

Figure A-1. Male pipe thread

Pipe threads lock together mechanically as the tapered male and female threads begin to interfere with one another as they are assembled (Figure A-2).

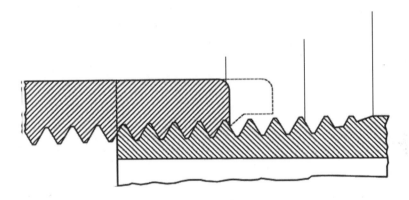

1. Overall thread length.
2. Length of good thread.
3. Length of imperfect thread (known as varnish threads on pipe, due to the chamfer on the thread die).
4. Length of hand-tight engagement.
5. Fully engaged position: tighten the pipe thread joint until the threads are engaged by the nominal engagement distance shown in Column 7 (see Table A-1). Should this not be possible, use the following guidelines for engagement:
   • Greater than hand-tight as shown in Column 5.
   • No more than the length of good threads as shown in Column 6.

Table A-1. National Pipe Thread data

1	2	3	4*	5	6*	7
Nominal Pipe Size, Inches	Pipe OD, Inches	Threads per Inch	Overall Thread Length, Inches	Hand-Tight Engagement (Inches/ Threads)	Length of Good Threads (Inches/ Threads)	Nominal Engagement for Tight Joint
1/16	0.3125"	27	0.390"	0.160"/4.3	0.261"/7.1	1/4"
1/8	0.405"	27	0.392"	0.162"/4.4	0.264"/7.1	1/4"
1/4	0.540"	18	0.595"	0.228"/4.1	0.402"/7.2	3/8"
3/8	0.675"	18	0.601"	0.240"/4.3	0.408"/7.3	3/8"
1/2	0.840"	14	0.782"	0.320"/4.5	0.534"/7.5	1/2"
3/4	1.050"	14	0.794"	0.339"/4.8	0.546"/7.6	1/2"
1	1.315"	11.5	0.985"	0.400"/4.6	0.683"/7.9	5/8"
1-1/4	1.660"	11.5	1.009"	0.420"/4.8	0.707"/8.1	11/16"
1-1/2	1.900"	11.5	1.025"	0.420"/4.8	0.724"/8.3	11/16"
2	2.375"	11.5	1.058"	0.436"/5.0	0.757"/8.7	3/4"
2-1/2	2.875"	8	1.571"	0.682"/5.5	1.138"/9.1	15/16"
3	3.500"	8	1.634"	0.766"/6.1	1.200"/9.6	1"

*Machined fittings, such as the one pictured in Figure A-3, have no varnish thread. The total thread length is also somewhat shorter than the overall thread length given in the pipe thread table, and the length of good thread is slightly longer. However, full engagement of machined pipe thread fittings is still the same as for other pipe threads.

Total thread length (all good threads)

Figure A-3. A pipe thread on a machined fitting is only slightly different than one on the end of a pipe

What Maintenance Technicians Should Know about Pipe Threads

1. Pipe threads are made to National Pipe Thread (NPT) Standards, as indicated in Table A-1.
2. Pipe threads are referred to by size, threads per inch, and an identity signifying a tapered thread:
   - MPT specifies a male pipe thread.
   - FPT specifies a female pipe thread.
   - 1/2 – 14 NPT specifies any 1/2″ national pipe thread (either male or female).
   - 1/4 – 18 MPT specifies a 1/4″ male pipe thread.
3. Pipe threads are sealed by filling the void area between their threads with a material such as Teflon tape or any number of pipe joint compounds made for this purpose.
4. To use Teflon tape:
   - Wrap the tape around the male thread clockwise when looking into the threaded end of the fitting or pipe. Wrap a total thickness of 6 mills of tape. For example, this is two wraps of 3-mill tape or four wraps of 1-1/2-mill tape. The wraps may also have to be spiraled onto the thread if the tape width is less than the engagement area of the threads. Cover the entire engagement area of the male pipe thread with six mills of Teflon tape (Figure A-4).
   - Pull the tape fairly tight as it is wrapped around the threads to get it into the root of the threads.

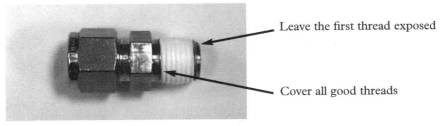

Leave the first thread exposed

Cover all good threads

Figure A-4. Teflon tape wrapped on a pipe thread

- Do not cover the first thread with tape—the one at the end of the pipe. This will prevent tape from getting into the pipe joint when it is assembled and contaminating the system. (Many pressure regulators and sensitive pieces of pneumatic equipment have become plugged with small pieces of Teflon tape because the first thread of the male pipe fitting was not left bare of tape.)
- Assemble the threaded joint to the correct engagement dimension (see Table A-1).

5. To use pipe joint compounds:
   - Select a joint compound that is compatible with the fluid being used in the system.
   - Cover the male pipe thread with joint compound over the entire engagement area of the thread. Leave the first thread bare.
   - Push the joint compound into the root of the threads and assemble the joint to the correct engagement dimension (see Table A-1).

Pipe Threads on Hoses

The most common type of pipe thread found on the end of a pre-made hose is a male pipe thread. Figure A-5 shows a typical hydraulic hose that can be purchased in various diameters and lengths at almost any industrial hardware supply.

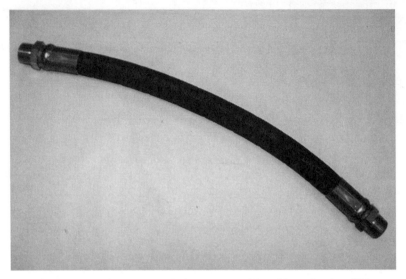

Figure A-5. A typical pre-made hose assembly with male pipe fittings

However, this type of hose would have an installation problem were it not provided with a special feature. The fittings on the hose ends are both solid and do not swivel. Once one end of the hose is screwed into its fitting on the machine, how can the other end then be screwed into its fitting without rotating the hose and unscrewing the original connection?

The solution to this problem is a special seat on the end of each male pipe thread. This seat is typically found only on male NPT hose fittings. The end of the male pipe thread contains a 30° seat, as shown in Figure A-6.

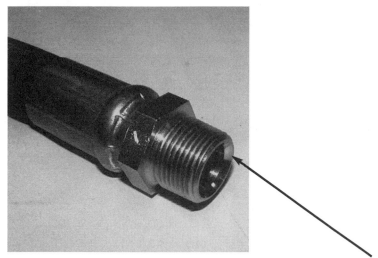

Figure A-6. Note the 30° (60° included angle) female seat on the end of this hose's male pipe thread

Note how this seat is unique to pipe threads on the ends of hoses. Compare it to the plain end of a male pipe thread found on a typical pipe fitting—shown in Figure A-7.

Figure A-7. A typical male pipe thread fitting. Note the plain end without any type of seat

The special seat on hose male pipe threads has been supplied to mate with *swivel pipe fittings*.

- Swivel pipe fittings contain a female nut that swivels before it is tightened. (It will not swivel after it is tightened.)
- Swivel pipe fittings contain a male seat that mates with the female seat on the end of the hose fitting.
- The swivel pipe fitting allows the hose to be attached to a machine without having to rotate the hose to screw it into a connecting fitting.
- A swivel pipe fitting is a seat-sealing fitting. It does not use sealant of any kind in the threads. The threads are only used to pull the seats together.
- Although the male pipe thread is a tapered pipe thread, the female thread in the swivel fitting is a straight thread. It is often called a "straight pipe thread" because it has the same diameter and thread pitch as a pipe thread, but no taper. They will screw together but will never begin to interfere with each other like a tapered male and female thread. The "straight pipe thread" on a swivel pipe fitting is called an American Standard Unified Thread (Figure A-8).

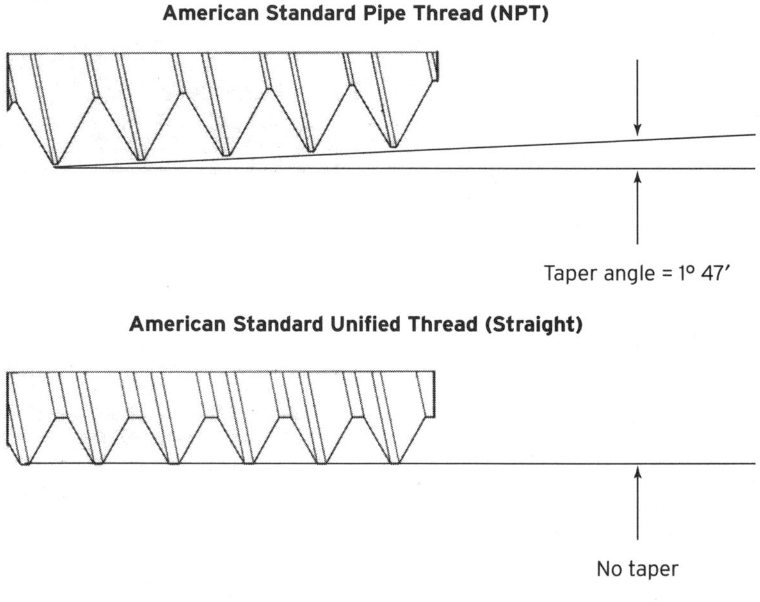

American Standard Pipe Thread (NPT)

Taper angle = 1° 47′

American Standard Unified Thread (Straight)

No taper

Figure A-8. Two American Standard thread patterns

Figure A-9 shows a typical male pipe hose end with a swivel adapter, disassembled.

Figure A-9. Typical male hose fitting with its mating swivel adapter

Figure A-10 shows the above fittings assembled. They are assembled by turning the swivel nut, which allows the hose to remain stationary during assembly.

Figure A-10. An assembled pipe swivel adapter on the end of a hose–sealed by the seats and not by their threads

A wide variety of swivel pipe adapters are available to suit most needs (Figure A-11).

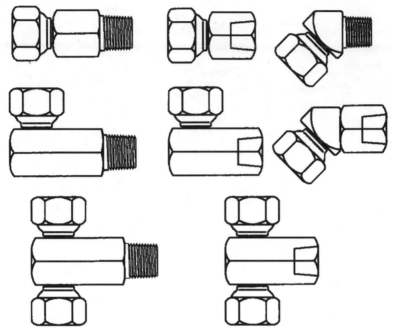

Figure A-11. A Variety of swivel pipe adapters

EXERCISES

1. Practice wrapping different sizes of male pipe threads with Teflon tape. Try different tape thicknesses (e.g., 1-1/2 mill and 3 mill tape) and different widths of tape (e.g., 1/8″ and 1/4″). Pay careful attention to leaving the leading thread bare and covering the entire length of good threads. Practice on pipe ends as well as machined fittings.

2. Practice assembling several different sizes of threaded pipe joints. Try combining different materials like inserting a steel male thread into a PVC female pipe thread. Assemble joints with both Teflon tape and pipe joint compound. Pay attention to the proper engagement length when tightening these joints.

3. Inspect the varnish threads on the end of a threaded pipe. Compare these threads to the same size pipe thread on a machined fitting and notice the difference.

4. Examine the male pipe threads on the end of hoses supplied for the class and compare them to male pipe threads on the ends of typical male pipe fittings. Notice the female seat?

5. Assemble male pipe hose ends to various swivel connectors. Note how they assemble without the threads interfering as they do on a typical tapered pipe joint. Notice how they produce a leak-tight seated joint. Does any Teflon tape or pipe joint compound need to be used on the threads of swivel pipe fittings? Why or why not?

JIC (AN) FITTINGS

JIC fittings are used primarily on hose ends and are often denoted as JIC (AN) fittings. The (AN) refers to Army/Navy. This type of fitting was originally developed as a standard by the U.S. government for military equipment. JIC fittings serve the same purpose on the ends of hoses as do swivel pipe fittings—they allow the hose to be installed without having to rotate it. However, JIC fittings differ from pipe swivel fittings in several ways:

- All of the JIC fittings contain matching straight threads, unlike a swivel pipe fitting, which mates a straight female thread to a tapered male thread.
- The seats in JIC fittings are larger than the seats in pipe swivel fittings. The angle of the JIC seat is 37° (Figure A-12).

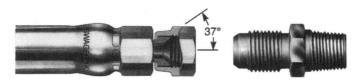

Figure A-12. JIC fitting seat detail

- JIC fittings always have a female seat included in the female nut and a male seat on the end of the male thread. The female nut always swivels before assembly. (It will not swivel after it is tightened.)

Figure A-13 provides a better look into a typical JIC fitting.

Figure A-13. JIC fitting, disassembled

Table A-2 identifies the threads used in common sizes of JIC fittings.

Table A-2.

JIC Fitting Size	Nominal Hose Size	JIC Thread Size
2	1/8"	5/16"–24
4	1/4"	7/16"–20
6	3/8"	9/16"–18
8	1/2"	3/4"–16
10	5/8"	7/8"-14
12	3/4"	1-1/16"-12
14	7/8"	1-3/16"-12
16	1"	1-5/16"-12
20	1-1/4"	1-5/8"-12
24	1-1/2"	1-7/8"-12
32	2"	2-1/2"-12

EXERCISES

1. Examine and assemble several of the different size JIC fittings supplied to the class.

2. Try to mate JIC fittings with the seat on the end of a hose containing a male pipe thread. How are JIC and swivel pipe fittings different from each other?

SAE FITTINGS (STRAIGHT THREAD–O RING BOSS)

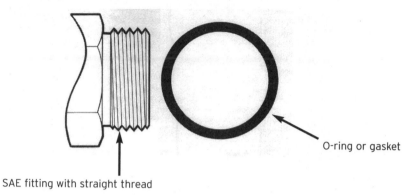

O-ring or gasket

SAE fitting with straight thread

Figure A-14. Typical SAE O-ring boss fitting, exploded

An SAE fitting is a straight male threaded fitting with an O-ring boss that screws into a thread of the same size on the mating part (Figure A-14). The seal is made by the O-ring or gasket used between the two parts. Shown in Figure A-15 are both a drawing and a picture of an SAE fitting with its O-ring in place. SAE thread sizes are incompatible with NPT threads.

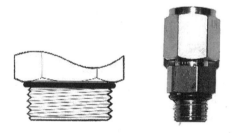

Figure A-15. Typical SAE O-ring boss fitting, assembled

The straight machine threads on the SAE fitting screw into the female straight thread with no resistance—just like a bolt screwing into a nut. The fitting pulls tight when the O-ring makes full contact with both parts, compressing the O-ring. The O-ring makes a leak-tight seal between the mating parts—not the threads.

Figure A-16. Typical SAE O-ring boss fitting, assembled

SAE Fitting Sizes

Table A-3 identifies the different sizes of SAE fittings commonly used.

Table A-3.

SAE Fitting Size	Thread Size
2	5/16-24
3	3/8-24
4	7/16-20
5	1/2-20
6	9/16-18
8	3/4-16
10	7/8-14
12	1-1/16-12
14	1-3/16-12
16	1-5/16-12
20	1-5/8–12
24	1-7/8–12
32	2-1/2-12

EXERCISES

1. Assemble various sizes of SAE fittings.

2. Practice tightening these fittings to different levels of torque and observe the different compression of the O-ring. What can the O-ring tell you about the correct amount of tightening torque?

3. Use a magnifying glass to examine the boss surfaces on both the SAE fitting and the port into which it is being inserted. Note the smooth flat sealing surfaces.

FITTING ADAPTERS

A wide variety of fitting adapters are available that convert pipe, SAE, and JIC fittings from one style to another. They come in a variety of combinations and sizes. Some are straight, while others are 45° or 90° elbows. They are available in any combination of genders. They also come in a wide variety of materials, such as steel, stainless steel, and brass. Most pipe-to-pipe fittings are also available in PVC.

The following are some common types of readily available fitting adapters.

- Pipe to pipe
- JIC to JIC
- JIC to pipe
- SAE to pipe
- SAE to JIC

Examples are shown in Figure A-17.

| SAE Boss to | SAE Boss to | Male JIC to |
| Male JIC | Female Pipe | Male Pipe |

Figure A-17. Fitting adapters are available in a wide combination of varieties

SWAGELOK TUBE FITTINGS

The following material applies only to Swagelok-brand tubing fittings used in low-pressure service with nonmetallic tubing. Low-pressure service for a Swagelok fitting is pressure below 1000 psi, so the pressure rating of the assembled tubing system will be the pressure rating of the plastic tubing. Any plastic tubing—such as polyethylene, nylon, vinyl, PVC, and Teflon—can be used with a Swagelok tube fitting.

The Swagelok tube fitting consists of a body, nut, front ferrule, back ferrule, and of course, the tubing (Figure A-18). These components are not compatible with similar components from other brands of tube fittings. There is no universal standard for these types of fittings, and parts from another manufacturer cannot be used interchangeably with Swagelok parts.

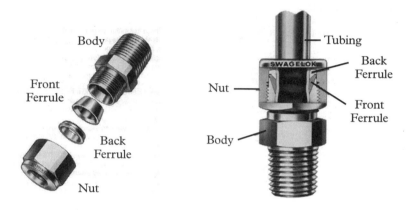

Figure A-18. Swagelok tube fittings, exploded & cross-section

When the fitting is properly assembled, the front ferrule, back ferrule, and tube will be permanently swaged together. This swaging process is shown in Figure A-19.

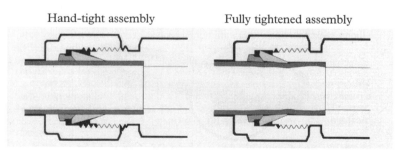

Figure A-19. The ferrules are swaged onto the tube when the fitting is first assembled

When very soft plastic tubing, such as Tygon or unplasticized PVC tubing, is used in a Swagelok fitting, an internal tube support is required so the ferrules can swage onto the tube. Tubing is considered to be soft enough to require a tube fitting if you can easily pinch the tube end together between your thumb and finger. If this is very difficult to do—leaving tube impressions in your fingers—then the tubing does not require an internal tube support (Figures A-20 and A-21).

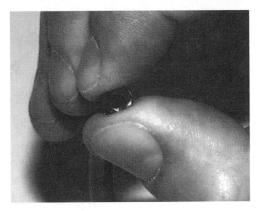

Figure A-20. This tube cannot be compressed easily, so it does not require a tube support

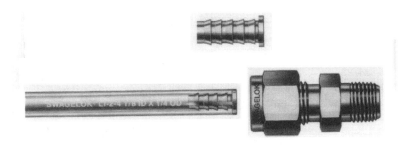

Figure A-21. Swagelok internal tubing support

To assemble a Swagelok fitting onto a tube for the first time:

1. Slip the nut, back ferrule, and front ferrule over the tube in the proper order and direction. Use a tubing insert if one is required.
2. Hold the Swagelok body secure in a vise if it is not already firmly attached to part of a machine.
3. Cut the end of the tubing square.
4. Insert the tube fully into the Swagelok body. Make sure that the tubing rests firmly against the shoulder inside the fitting.
5. Tighten the nut onto the body hand-tight.
6. Tighten the nut an additional 1-1/4 turns for tubing 1/4" OD and greater. (Tighten the nut an additional 3/4 turns for 1/16", 1/8" and 3/16" tubing.)

   (Note: It is a good idea to put a mark on the nut and body when they are hand-tight to help you keep track of the number of turns made to tighten the nut.)

To reassemble a Swagelok fitting that has already been used.

1. Check that the front and back ferrules have clearance between them. You should be able to insert a fingernail between them. If not, the ferrules have been overtightened and no longer will function properly. They must be cut off the end of the tubing and replaced with new ones (Figure A-22).

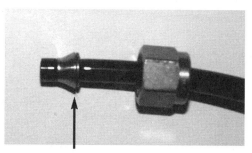

Figure A-22. Because of overtightening, these ferrules have no space between them and must be replaced

2. Inspect the bevel in the body of the fitting. Replace the Swagelok body if any damage is found:
   • The top lip should not be enlarged or rolled outward—you can compare the fitting to a new one to see if there is any noticeable enlargement.

- The bevel area where the ferrules seat should be smooth and free of any noticeable scratches or pitting.

3. Reassemble the tube assembly to the body. Tighten the nut hand-tight. Then tighten the nut an additional (1) flat to secure the fitting to a leak-tight state (Figure A-23).

Re-assembly of a used Swagelok fitting—turn 1 flat past hand-tight

Figure A-23. Re-assembly sequence of used Swagelok fittings

Inspecting Assembled Swagelok Tube Fittings

Assembled Swagelok fittings can be inspected for certain deficiencies.

There should be one thread showing beyond the tube nut. If more than one thread is showing, the fitting may be undertightened or some other problem with the assembly exists. If less than one thread is showing, the Swagelok fitting has been overtightened.

(Note: An overtightened fitting will most often seal the first time that it is assembled, so long as it is not severely overtightened. However, if loosened and retightened, the fitting will almost surely start to leak. The ferrules—and possibly the body—will have to be replaced to correct the damage done by overtightening.)

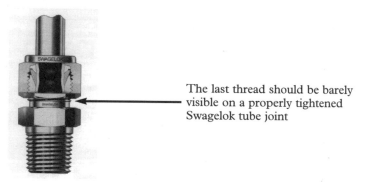

The last thread should be barely visible on a properly tightened Swagelok tube joint

Figure A-24. Inspecting an assembled Swagelok tube fitting

EXERCISES

1. Assemble a Swagelok fitting onto a 1/4″ plastic tube hand-tight, and then tighten the nut 1/4 turn. Disassemble the fitting and observe what has happened to the ferrules. Re-assemble the fitting hand-tight and tighten the fitting another 1/4 turn. Disassemble the fitting again and observe the changes to the ferrules and tube. Repeat this exercise, tightening the nut an additional 1/4 turn each time until it has been turned 1-1/4 turns. Now observe the ferrules and the properly assembled tube fitting. What can you learn about how a Swagelok fitting seals by these observations?

2. Continue to tighten the Swagelok fitting onto the tube 1/4 turn at a time. You will now be over-tightening the fitting. Observe what happens to the ferrules as they are overtightened. How many extra 1/4 turns does it take to remove the spacing between the front and rear ferrules on a 1/4″ Swagelok fitting?

3. Try assembling different tube materials to a Swagelok fitting. Of the tubing made available to the class, which ones need to have tubing inserts in their ends before swaging the ferrules?

APPENDIX B

Answers to Questions About Fasteners

These are examples of the types of questions to investigate when developing a maintenance training course on basic machine components (in this case, fasteners).[1]

PHILLIPS AND POZIDRIVE SCREWDRIVERS

Agilent toolboxes in the IC fab include both Phillips and Pozidrive screwdrivers. They appear to be very similar, as shown in Figure B-1; the top driver is a #2 Pozidrive, while the lower driver is a #2 Phillips.

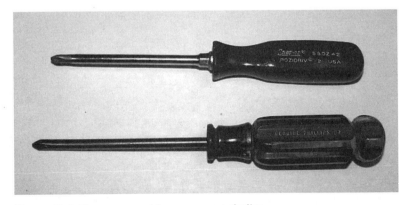

Figure B-1. These screwdrivers appear similar

---

[1]These questions were asked in the Chapter 6 subsection, "Preparing Technical Minicourses."

Questions: What is the difference between a Phillips and a Pozidrive screwdriver? Are they interchangeable? If not, how do you know when to use one or the other?

Answers: Both Phillips and Pozidrive screwdrivers are cross-recessed drivers. However, the Phillips recess has tapered walls, while the Pozidrive recess has straight walls. They are not interchangeable. Using the incorrect screwdriver in a cross-recessed screw head will likely damage the screw. A Phillips screwdriver must be used in a Phillips screw, and a Pozidrive screwdriver in a Pozidrive screw. The marking on a Pozidrive screw head distinguishes them (Figure B-2).

Phillips Head Screw   Pozidrive Head Screw

Figure B-2. Phillips and Pozidrive screws

Questions: On Phillips head screws, when do you use a #1 Phillips? A #2 Phillips? A #3 Phillips?

Answers: Most fasteners are manufactured to standards that provide for the following sizes of Phillips screwdrivers.

Screw Size	Phillips Screwdriver Size
#4	#1
#6 through #10	#2
Larger than #10	#3

HEX HEAD AND SOCKET HEAD BOLTS

Different kinds of hex head bolts are stocked in Agilent's spare parts bins.

Question: What do the different marks on the top of these hex head bolts mean? (See Figure B-3.)

No Lines   3 Lines   5 Lines

Figure B-3. Hex head bolts with different markings on their heads

Answers: Marks on the tops of hex head bolts indicate material properties of the bolt. The most common markings for English bolts are shown in Figure B-4.

SAE Grade 2   SAE Grade 5   SAE Grade 8

SAE Grade	Material Tensile Strength
#2 (Butter bolt)	75,000 psi
#5 (Common production machine bolt)	120,000 psi
#8 (High-strength bolt)	150,000 psi

Figure B-4. Common markings for English bolts

Socket head bolts are also stocked but have no such markings on their heads.

Question: How do socket head bolts compare in strength to different kinds of hex head bolts?

Answer: Socket head bolts are most commonly black (unfinished), zinc-plated, or stainless steel (Figure B-5). Black and zinc-plated socket head bolts generally have a tensile strength of 170,000 to 190,000 psi, making them harder and stronger than SAE Grade 8 bolts.

However, stainless steel bolts, despite their reputation for being hard, are little better than butter bolts as far as strength is concerned. They have a tensile strength of 100,000 psi, but their yield strength is only about 65 percent of tensile, compared to yields of 80 percent of tensile in nonstainless ferrous fasteners. This makes stainless steel bolts very similar in useable strength to SAE 2 bolts.

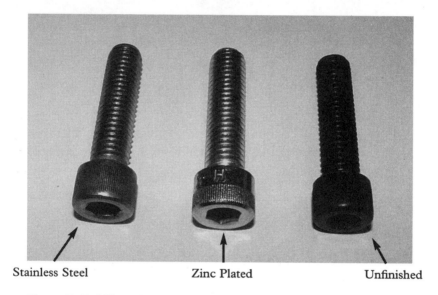

Stainless Steel   Zinc Plated   Unfinished

Figure B-5. Different kinds of socket head bolts

LOCK WASHERS

The lock washers found in Agilent's hardware stock bin are shown in Figure B-6.

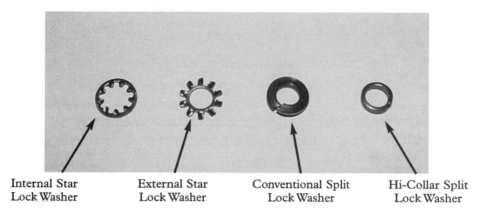

| Internal Star Lock Washer | External Star Lock Washer | Conventional Split Lock Washer | Hi-Collar Split Lock Washer |

Figure B-6. Different kinds of lock washers

Questions: When do you use a split lock washer versus a star lock washer? Why do some split lock washers look different from others?

Answers: Split lock washers are the most common locking devices used for threaded fasteners. Hi-collar split lock washers are most commonly used with socket head bolts because of their extremely small diameter head, especially when these bolts are placed in machine counter bores.

Star lock washers lock fasteners more securely than split lock washers and create joints with better electrical conductivity. However, they generally cause more damage to the mating surfaces of the fastened joint.

Question: What is the difference between external star and internal star lock washers?

Answer: External star and internal star lock washers are used depending on the size of the bolt head and the size of the clearance hole provided for the bolt. Socket head bolts can only be used with an internal star washer—the "stars" on an external star washer do not even come into contact with the head of a socket head bolt. Hex head bolts may use either, but bolts in large clearance holes will require an external star washer.

OTHER QUESTIONS

Other questions our technicians asked about threaded fasteners and their use included the following.

Question: What do I have to know in order to specify a bolt completely?

Answer: To specify a bolt, the following information is needed:

Bolt Property	Examples
Diameter	#8, 1/4"
Thread pitch	20, 32, 40 threads per inch
Thread standard	UNC, UNF
Thread class	1A, 2A, 3A
Thread specialties	Full-length thread, left hand
Material properties	SAE 2, SAE 8, stainless steel
Head type	Flat head, hex head, socket head
Recess (if used)	Phillips, Pozidrive, hex key, Torx

Example:
1/4"—20—UNC—3A—full-thread—SAE 5—hex head

Question: For a standard nut-and-bolt assembly, what is the weak link—the bolt shank, the bolt thread, or the nut?

Answer: A properly designed bolt and nut assembly should never fail by stripping the thread on either the bolt or nut. The bolt shank should always fail at the minor diameter of a thread above the nut.

A standard nut provides enough threads to prevent thread stripping—a heavy nut even more. Jam nuts do not contain enough threads to provide such strength. Jam nuts were designed only to "jam" against a standard nut to form a locking device.

Question: Which are stronger—coarse or fine threads?

Answer: Eighty percent of the strength developed in a threaded joint is provided by the first three complete threads, and virtually 100 percent of the strength is taken by the first complete five threads.

A fine thread makes a stronger bolt, because the fine thread has a larger minor diameter that provides a larger surface area for the bolt shank at its designed point of failure.

Nuts, however, are stronger with a coarse thread. Nuts can only fail by shearing the thread at its root, and a coarse thread has a larger shear area at its root than a fine thread. Since the first three threads carry most of the load, it is somewhat inconsequential that a fine-threaded nut contains more threads than its counterpart with a coarse thread.

Question: How do I know how much to tighten a bolt?

Answer: Typically, threaded fasteners are tightened to torque values taken from torque charts, given the diameter and material properties of the fastener. For situations where standard torque values do not apply, create a test piece and break five to ten fasteners. Set the tightening torque value to 70 percent of the minimum torque required to break these samples.

Glossary of Agilent Manufacturing Terms

Clean room	The primary manufacturing area in an IC fab, which maintains a circulation of extremely clean air to control airborne contamination. (Small particles destroy the micro-electrical circuits being manufactured on an IC wafer.)
HAZMAT	Agilent's Hazardous Material Teams. These teams are trained to respond to incidents involving hazardous materials and to human health emergencies.
IC	Integrated circuit. An electrical device (often called a "computer chip") that is manufactured in an IC fab.
IC fab	An integrated circuit fabrication plant. A factory, containing a clean room, that produces integrated circuits.
IC wafer	A flat, round silicone disc on which multiple IC devices are manufactured.
Lockout/Tagout	OSHA regulations regarding safety procedures for locking and tagging a machine's electrical and other power sources (e.g., pneumatic air supply) so that it is safe for tech maintenance.
OSHA	The Occupational Safety and Health Administration, a U.S. government agency within the Department of Labor that oversees health and safety regulations for U.S. industries.
PLC	Programmable logic controller, a small, inexpensive computer, capable of connecting to many different types of hardware inputs and outputs. PLCs are often used to make customized machine controllers.

PM

A preventive maintenance procedure performed by equipment technicians to restore equipment deterioration and prevent machine failures.

Process recipe

A set of instructions entered into a machine's control system that instructs it in how to process a certain product. In an IC fab, machines make many different products, so they have many different process recipes loaded into their control system.

RF

Radio frequency electrical power. IC fabs use RF power supplies that produce alternating electrical current whose frequency is in the radio spectrum, rather than the 60-cycle-per-second frequency of standard electrical power.

Wafer lot

A box of 25 identical IC wafers, manufactured as a batch (all the wafers being the same product).

WIP

Work in progress, denoting material on a manufacturing line that has begun the manufacturing process but is not yet completed.

Index

1-point lessons. *See* One-point lessons
3S. *See* Sort Stabilize Shine
4M. *See* Man Machine Material Methods
5S principles. *See* Sort Stabilize Shine
    Standardize Sustain
5-why analysis. *See* Five-why analysis
7S. *See* Sort Stabilize Shine Standardize
    Sustain Safety Security

A

Abnormalities. *See* Wafer
    causation, 223
    restoration, 192, 193
Abrasive mixture, 213
Accelerated deteroriation, 23
Acceleration/braking, 224
Access. *See* Cleaning; Inspection
    areas, 89
    identification. *See* Machine
    improvements. *See* Maintenance
    panels. *See* Agilent
Access/cleaning, machine preparation, 70
Action teams, 33–34. *See also* Cross-
    departmental action teams
    implementation. *See* Total Productive
    Maintenance
Activity
    boards. *See* Agilent; Behavior Safety
        Awareness is Fundamental;
        Integrated circuit; Team
    process, 108
    checklist. *See* Total Productive
        Maintenance
    daily scheduling, 101–104

Actual product time, 271
Adhesives, 138
Adjustment
    analysis, 216–218
    setting, 217
    time, 229–231. *See also* Machine
Administrative solutions, 203
Advanced levels. See Machine failures
Agilent
    chiller, 237
    clean room, 76, 77, 215
    equipment, 51
    cleaning process, 234
    IC fab, 16, 135
    industry-specific tools, 139
    machine, 233, 244
    access panels, 100
    part, 122
    production machines, 196
    Three Lists examples, 71
    quality monitor checklist, 128
    safety program, 63
    TPM team activity board, 298
    vision/strategy/tactical plan,
        creation, 254
    visual factory floor, 158
Agilent manufacturing
    strategy, 254
    tactical plan, 254
    terms, 341–342
    vision, 254
Aging paradigms. *See* Equipment
Air leaks, 52
Aluminum thread, 187

American Standard thread patterns, 316
AN fittings, 137, 320–321
   exercises, 321
Analysis. *See* Adjustment analysis;
   Calibration analysis; Condition-of-
   use analysis; Failure; Life analysis;
   Lubrication analysis; Machine-part
   analysis; Maintenance; Physical
   analysis; Preventive maintenance;
   Productivity; Quality
Antecedent programs, 293
Antecedents, 289. *See also* Behavior
Assembled Swagelok tube fittings
   exercises, 331
   inspection, 330–331
Associate members, 38
Audits, 143, 156. *See also* Cleaning;
   Inspection; Manager; Safety
Availability losses, 268
Availability rate, 269, 271, 272

B

Bar-code card, 88
Bar-code readers, 231
Basic product time, 271–273
Bearings, 138
   covers/spacers, 150
   flange side, 149
   freezing, 189
Before-and-after data, comparison, 203
Before-and-after equipment
   photographs, 89
Before-and-after photographs, 95
Behavior. *See* Employee
   antecedents, 290
   basics, 289–292
   consequences, 290–292
   management, 287
   observation, 292
   results, 293
   safety
      committee, 67
      compliance, 68
   shaping, 293
Behavior Safety Awareness is
   Fundamental (BSAF)

Behavioral Safety Program Activity
   Board, 69
   program, 68
Behavioral safety
   goals, 68
   observer, training, 68
Bellows
   assembly. *See* Vacuum bellows
      assembly
   conditions-of-use, providing failure,
      190
   life, predictability, 190–191
Belts/pulleys, 138
Bolts, 52, 55, 162. *See also* Hex head
   bolts; Nut-and-bolt assembly;
   Socket head bolts; SS bolt
   clearance hole, 187
   failure. *See* Test bolts
   loosening, reasons, 186
   markings. *See* English bolts
   property, 338
   tightening, 143
   torque
      measurement, 188
      tests, 187
   torquing, 84
Bottleneck machines, 17. *See also*
   Non-bottleneck machines
   material flow rate, measurement, 282
   wafer capacity chart, 279
Breakdown maintenance, 28, 179, 207
   plan, 122
Breakdowns. *See* Machine
   fixing, 29
BSAF. *See* Behavior Safety Awareness is
   Fundamental
Built-in testing, 121
Business gains, 9
Business results, improvement, 8

C

Calibration/adjustment analysis, 216–218
Capacity assurance. *See* Machine
Cassette. *See* Wafer
   lift assembly, 193
   testing fixture, 240
Catwalk, addition, 74

Caulks, 138
Cause-and-effect diagrams, 202
Celebrations, 293
Ceramics, 138
Certification. *See* Safety; Training
    completion, 143
    requirements, 143
Chains/sprockets, 138
Chamber lift assemblies, 193
Change, leading, 249
Check sheets, 202
Checklist. *See* Equipment restoration;
    Maintenance plans; Maintenance
    plans implementation; Preventive
    Maintenance
    usage, 130
Chemical/gas systems, 139
Chronic conditions, TPM pyramid,
    24–27
Chronic loss, 22
C&I. *See* Cleaning and inspection
Circuit breaker, 238
Clean room, 341. *See also* Agilent
equipment. *See* Agilent
Cleaning. *See* Access/cleaning; Mindless
    cleaning
    access, 95
    audits, 105
    cart, 75
    cycle, 233
    improvements, 97–98
    maintenance work, 75
    materials, 63, 74–75
    purposes, 51
    standards, 63, 89–94
        changes. *See* Inspection; Operators
        creation, 47
    supplies, 75, 168
    tools, improvement, 98
    usage. See Machine defects
Cleaning and inspection (C&I)
    activities, decrease, 98
    standards, 63, 90–94, 102–104, 108, 170
    work order, 102–104
Clean-room garments, 178
Closed M-Tags, 78
Clutches/brakes, 138

Clutter, minor defects, 60
Color. *See* Teams
codes. *See* Lubricants
coding. *See* Production
    usage, 86
Color-coded nests, 159
Components. *See* Fluid transmission
    components
    deterioration. *See* Machine
    inspection. *See* Machine
    maintenance. *See* Machine
    selection, 28
    standards, 48–51
Compressor, noise, 59
Compressors, 138
Condition
    maintenance, 18, 19
    monitoring. *See* Continuous condition
        monitoring; Extended condition
        monitoring
Condition-based maintenance, 62
    plans, creation, 192
Condition-measurement trend
    charts, 117
Condition-of-use analysis, 222–228
Conditions-of-use. *See* Machine
    components
    providing, failure. *See* Bellows
    requirements, 223
    usage, 189
Connectors, looseness/deterioration,
    52, 59
Consequences, 289, 291, 292. *See also*
    Behavior; Extinction consequence;
    Punishing consequence
    delivery, 294
Constraints, concept, 16–17
Contamination, 52
    areas, 89
    containing, 98
    control activities, 168
    elimination, 98
    removal, 51
    sources, 95
Contamination, effect. *See* Defects
Contingent consequences, delivery.
    See Team members

Continuous condition monitoring, 241–243
Continuous improvement, development, 246
Continuous learning, 6
Continuous productivity, improvement, 7–9
Continuous-monitoring devices, 243
Control charts, 202
Control valves. *See* Cooling water; Pneumatic control valves
Controlled part stock, 164–166
Controlling hardware, 219–220
Controlling parameter, 219–220
Cooling loops, 225
Cooling manifold ports, 200
Cooling supply loop, 226
Cooling water
  control valve, 199
  lines, 198
  pressure, 225
  supply, 226
Core members, 38
Cost analysis. *See* Maintenance
Cotter pins, 55
Countermeasures, 298
  plans. *See* Maintenance plans implementation
Couplings, 138
Course. *See* Minicourses
  contents, 309
C-rings, 138
Cross-departmental action teams, 108
Cross-departmental focus teams, 248
Cross-departmental teams, 6
  deployment, 37
Cross-reference data, 163
Cryogenics, 139
Culture
  change. *See* Total Productive Maintenance
  change/resistance. *See* Maintenance
  contrast, 9
CVD Chamber Isolation Valve PM, 114, 116
CVD Chamber Throttle Valve PM, 114, 115

CVD chambers, 262
Cycle time, 201

D

Day-to-day reliability, 229
Day-to-day work routine, 64
Defects, 47. *See also* Equipment; Machine defects; Major defects; Medium defects; Minor defects
  contamination, effect, 56
  density, 202
  effect. *See* Machine failures
  foundation, 27
  rate. *See* Product
  spotting, 53
Deficencies, 68
Deliverables. *See* Equipment restoration; Machine failures; Machine productivity improvement; Maintenance plans; Maintenance plans implementation
Delta connections, 138
Design
  engineers, 20
  improvement projects, 161
Detector circuit, 242
Deterioration, 20. *See also* Accelerated deterioration; Forced deterioration; Graceful deterioration; Machine; Natural deterioration; Nongraceful deterioration
Dirt, 52
  elimination, 156
Disassembled robot, 146
Documentation. *See* Precision
  system. *See* Visual maintenance documentation system
  waste, 149
Doing. *See* Learning
Down machine, 180
Downtime, 271
  duration, 275
Drive belt, 56
Duration. *See* Downtime; Failure; Machine failures
Dust, elimination, 156
Dynamic seals, 137

E

Economic impact, 122–123
Electrical power transmission, 138
Electrical wiring, minor defects, 61
Electronics, 138
Employee
  behavior, 64, 67–68
  job security, 157
  work habits, 156
End results, 293
Engineered solutions, 203
Engineering change, 203
Engineers. *See* Production
  role, elevation, 171
English bolts, markings, 335
Equipment
  abnormalities, causes, 51
  aging paradigm, 17–18
  breakdowns, recognition, 211
  components, 28
    cleaning/inspection, 95
  defects. *See* Man-made equipment
    defect examples. *See* Minor
    equipment defects
    identification, 48
  design, 28
  failures, 26, 27, 207
    minor defects, 5
    rates, 246
    reduction, 297
    root cause, 27
    usage, 120
  goals, 47
  improvement, Pareto approach, 261
  labels, minor defects, 60
  losses, 49
    types, 23
  maintenance, 29
    goals, 18–21
    plans, 173
  minor defects. *See* Leaking
    equipment
  performance, 28, 53
  photographs. *See* Before-and-after
    equipment photographs
  productivity, world-class levels, 260
  redesign, development. *See* Machine

technicians, one-point lessons
    usage, 81, 82
Equipment restoration, 41, 45
  advance step, 96–105
  deliverables, 109
  goals, 47–48
  infrastructure support, 108–109
  master checklist, 105–107
  tookit, 63–82
Ergo loader, addition, 66
Evaluation systems, changes, 108
Extended condition monitoring, 234–243
Extinction consequence, 291

F

Factory
  day-to-day operation, 39
  fire/loss security measures, 156
  floor, 162, 248. *See also* Agilent
  building blocks, 282
  employees, 67
  losses, 248
  machine failure rate, reduction, 54
  output, 285
  productivity, increase, 246
Failure
  analysis, 179–205, 298. *See also*
    Preventive Maintenance
    process, 180
    tools, 184–204
  duration, 275
  location, 184
  object, 184
  occurrence, 196
  prediction, 189
  prevention
    occurrence, 189–190
    skills, 136, 144, 181
    trouble lists, 180–183
  process, 184, 185
  rate, reduction. See Factory
  timing, 184, 185
Fasteners, 52, 55, 137
Fasteners, questions/answers, 333
Feedback, providing, 294
Female pipe thread (FPT), 312
Female straight thread, 323

Ferrules, 326, 327
Filter-mounting device, 59
Filters, 83
Fishbones, 202
Fitting adapters, 325
Fittings. *See* AN fittings; Flanged fittings;
    Fluid fittings; JIC fittings; Male pipe
    fittings; Pipe threads; Pneumatic
    fittings; Quick-connect fitting;
    Quick-connect fittings; Swagelok
    tube fittings; Ulta-torr vacuum
    fittings; VCO fittings; VCR fittings
Five-why analysis, 184–189, 202
Five-wire systems, 138
Fixtures, 235, 240
Flange. *See* Bearings
Flanged fittings, 137
Flow charts, 202
Flow meter, visual controls, 84
Fluid fittings, 137
Fluid reservoirs, 52
Fluid transmission components, 137
Forced deterioration, 23
Four-wire systems, 138
FPT. *See* Female pipe thread
Frequency, 90
    interaction. *See* Shape

G

Gap analysis. *See* Total Productive
    Maintenance
Gap washer, 187
Gas analyzers, 139
Gas line leaks, 236
Gauges, 52, 57, 83. *See also* Pressure
    gauges; Temperature gauges
    marking material, 75
    minor defects, 57
    setpoints, 108
Gears/gear boxes, 138
Generic technology lessons, 79
GFI. *See* Ground-fault interrupt
Goals. *See* Equipment restoration;
    Machine failure; Machine
    productivity improvement;
    Maintenance plans; Maintenance
    plans implementation; Team

Graceful deterioration, 120–121
Gross vehicle weight (GVW), 223
Ground lugs, 61
Ground-fault interrupt (GFI), 121
Guards, addition, 73
Guided practice, providing, 11, 140
GVW. *See* Gross vehicle weight

H

Handling blade, 185
Hardware. *See* Controlling hardware;
    Hydraulic system hardware/
    schematics; Pneumatic system
    hardware/schematics
    maintenance. *See* Machine
Hazardous Material (HAZMAT)
    procedures, 68
    teams, 68, 341
HAZMAT. *See* Hazardous Material
Hex head bolts, 141, 335–336
High-G cornering, 224
High-voltage terminals, protection, 72
Histogram. *See* Machine failures
Histograms, 202
Hoses, 137, 185, 226
    assembly. *See* Pre-made hose
        assembly
    pipe threads usage, 314–318
    size, 321
HVAC tools, 139
Hydraulic system hardware/
    schematics, 137

I

IC. *See* Integrated circuit
Improvement
    activities
        focus, 8
        focusing. *See* Machine
    ideas. *See* Team
    metrics, 265
        data, 117
    number, 268
    projects. *See* Design
    teams, 39, 73
Index cards, 117
Indicator, change, 201

Infrared thermography, 235, 237–238
    tool, 238
Infrastructure support. *See* Equipment
    restoration; Machine failures;
    Machine productivity improvement;
    Maintenance plans; Maintenance
    plans implementation
Inherent life expectancy, 23
Input station. *See* Visually controlled
    input stationr
Inspection. *See* Visual inspections
    access, 95
    audits, 105
    cart, 75
    cleaning standards, changes, 205
    maintenance work, 75
    materials, 63, 74–75
    points, 85, 86
    routes, 85. *See also* Visually controlled
        inspection routes
    specifications, 113, 125
    standards, 63, 89–94
        creation, 47
    station number, 85
Integrated circuit (IC), 341
    fabrication, 68, 88, 212, 225, 341.
        See also Agilent
        personnel, 67
        Safety Activity Board, 69
    processing machines, 233
    wafer, 341
Internal filter screen, location, 200
Inventory, 167
    control computer, 164
I/O valve, 193

J
JIC fittings, 137, 320–321
    exercises, 321
    seat, 320
    size, 321
JIC thread size, 321
Jigs, 235, 240
Job security. *See* Employee
Joints. *See* Pipe joint
failure, 186, 188

K
Keys, 138
Knowledge
    elevation, 170–171
    lack, 49
    need, 79

L
Lapping pad, replacement, 244
Leaking equipment, minor defects, 58
Leap of faith, 96
Learn Use Teach Inspect (LUTI), 11
Learning. *See* Continuous learning
    doing, 11, 34, 140
    principles, 11
    time, 136, 144
Lessons. *See* Generic technology lessons;
        Machine-specific one-point lessons;
        One-point lessons
Life analysis, 222–228
Life expectancy, 21, 190. *See also*
        Inherent life expectancy; Natural
        life expectancy
    concept, 122
    curve, 122
Line disruption, 284
Line items, scheduling, 108
Linear product flow, 281–285
Liquid leaks, 52
Lists. *See* Three Lists
Locators. *See* Visual route maps
Lock washers, 142, 337
    types, 337
Locking pins/devices, 52
Lockout, 341
Loss. *See* Chronic loss; Sporadic loss
    types. *See* Equipment
Lot count, 283
Low-priority parts, 222
Lubricants
    color codes, 214
    procuring, 214
    storage, 212
    usage, 213, 215
    visual identification system, 214

Lubrication, 138
 analysis, 212–216
 maintenance, execution, 212
 needs, 213–214
 points, identification, 213
 scheduling system, creation, 212
LUTI. *See* Learn Use Teach Inspect

M

Machine
 abnormal sounds, 52
 access, 73
  identification, 70
  preparation, 70
 adjustment times, 268
 availability, 267
 breakdown, 28, 229, 268
  economic impact, 123
  Pareto charts, 259
 capacity
  assurance, 276–281
  data, 278
 cleaning, 95
  preparation, 70
 cleaning and inspection (C&I)
  standard, 91–94
 control system, 60
 controls. *See* Visual machine controls
 design. *See* Weak machine design
 deterioration, 22–23
 equipment redesign, development, 124
 factors, identification, 192, 193
 hardware, maintenance, 218
 improvement
  activities, focusing, 255
  ideas, 89
 maintenance, 28
  plan, development, 124
  scheduling plans, 117
 metrics, trend charts, 89
 minor stoppages, 267
 misapplication, 20
 mistreatment, 20
 motions, 52
 odors, 52
 operation, 201, 238
  ceasing, 185, 189

 restoration, 26
 optimal conditions, defining, 192, 193
 parts, 224, 261
 performance, 18, 19, 53–54
  improvement, 13
 physical analysis, 196
 preparation. *See* Access/cleaning;
  Production
 preventive maintenance plan,
  identification, 113–128
 productivity, 53
  improvement, 245
  loss, Pareto charts, 258
 redesigns, 123–124
 restoration, 47, 196
 service data. *See* Preventive
  Maintenance
 setup, 229–231
  times, 267
 speed, 268
  reduction, 229, 233–234
 standards setting, team members
  usage, 61–62
 stoppages, 229, 232, 268
 technologies, 136
 vibrations, 52
Machine components, 136, 222, 224
 conditions-of-use, 20
 deterioration, 49
 inspection, 51
 maintenance, TPM focus, 28–29
Machine defects, 25, 48–62, 95
 detection, cleaning usage, 51–53
 types, 48
Machine failures, 24, 25
 advanced level, 183–184
 defects effect, 48, 53–61
 deliverables, 207
 duration, 274–275
 histogram, 275
 infrastructure support, 207
 master checklist, 206
 number, 267, 274
 practiced levels, 182, 183
 prevention, 42, 175, 206
  goals, 177–205
 rate, reduction. *See* Factory

repairs, 183
root cause, 26, 109
start-up level, 181–183
types, 182
Machine loss, 22–23
TPM attack, process, 23–242
Machine productivity improvement,
42, 209
deliverables, 246
goals, 211–244
infrastructure support, 246
master checklist, 245
team approach, 33
TPM usage, 15–16
Machine-part analysis, 222
Machines
linking, 230
Machines skills, 143–144
Machine-specific one-point lessons, 79
Maintenance. *See* Condition
maintenance; Condition-based
maintenance; Precision; Reactive
maintenance; Replacement
maintenance; Routine maintenance;
Time-based maintenance; Total
Productive Maintenance; Use-based
maintenance
access improvements, 98–101
analysis, 184, 189–191. *See also* Quality
cost analysis, 244
culture, change/resistance, 10
departments, 173
documentation system. *See* Visual
maintenance documentation
system
goals. *See* Equipment
kit, 167
organizations, 207
procedure, 145, 297, 298. *See also* Text
maintenance procedure
description, 149
roles, elevation, 170–171
routines, 5–6
schedules
creation. *See* Restoration
design, 120–124
strategy, 33

support tools. *See* Precision
maintenance support tools
technicians, 109, 114, 135, 145, 233, 246
activity, 10
pipe thread knowledge, 312–313
technology, acquisition, 205
time, ratio, 268
tools, 166–169
organization, 169
TPM focus. *See* Machine component
maintenance
training material, example, 309
work, 134. *See also* Cleaning;
Inspection
scheduling, 120
Maintenance plans, 298. *See also*
Breakdown maintenance;
Technician-scheduled maintenance
plans; Underdeveloped
maintenance plans
assembling, 116
clarity, 117
creation, 192, 194. *See also* Condition-
based maintenance; Time-based
maintenance; Use-based maintenance
deliverables, 130
design, 220
development, 123
goals, 113–128
identification, 41, 111, 129. *See also*
Machine
infrastructure support, 130
master checklist, 129
usage, 189–190
Maintenance plans implementation,
42, 131
countermeasure plans, 205
deliverables, 173
goals, 133–171
infrastructure support, 173
master checklist, 171–172
precision execution, 136–169
technical skills, 136–143
timeliness/completion, 134–136
Maintenance Tags (M-Tags), 63, 76–78,
89, 95. *See also* Closed M-Tags;
Open M-Tags

example, 76, 77
number, 96
system, 108
Maintenance-tool storage system, 173
Major defects, 48
Male hose, 317
Male pipe fittings, 314
Male pipe thread (MPT), 312, 315
fitting, 315
Man Machine Material Methods (4M)
sources, 223
Management systems. *See* Safety
Managers, 156
audits, 108
Model Teams, 35, 96
Man-made equipment defect, 50
Manufactured product, quality, 218
Manufacturing
costs, 17
design, 28
process, 28, 232
design, changes, 205
productivity, 297
impact, 278
strategy. *See* Agilent manufacturing
vision. *See* Agilent manufacturing
Master checklist, 106–107. *See also*
Equipment restoration; Machine
failures; Maintenance plans;
Maintenance plans implementation
Master PM plan. *See* Production
Master precision maintenance, 139
Materials, 138
accumulation, 156
breaking down, 140
color coding. *See* Production
flow rate, measurement. *See* Bottleneck
machines
Mating swivel adapter, 317
Mean time to failure (MTTF), 267
Mean time to repair (MTTR), 267, 274
Measurement system, design/
implementation, 293
Mechanical drawings, 137
Medium defects, 48
Members. *See* Associate members; Core
members; Team

Metal dep, 71
Metrics. *See* Improvement; Machine
trend charts. *See* Machine; Team
Metrics, support, 293
Mindless cleaning, 95
Minicourses, preparation. *See* Technical
minicourses
Minor defects, 48, 53, 223. *See also*
Clutter; Electrical wiring;
Equipment; Gauges; Leaking
equipment; Visual controls
detection, 52
inspection, 237
Minor equipment defects, 109
detection, 59
examples, 55
Minute-to-minute reliability, 229
Mistake-proof designs/procedures, 205
Model Teams. *See* Manager
Monitoring. *See* Continuous condition
monitoring; Extended condition
monitoring
Monitors. *See* Schedule monitors
installation, 121
Motion controllers, 138
Motor-condition analysis, 235, 239
Motor-condition analyzer, 239
Motors, 138
life, reduction, 228
Move rates. *See* Product
MPT. *See* Male pipe thread
M-Tags. *See* Maintenance Tags
MTTF. *See* Mean time to failure
MTTR. *See* Mean time to repair

N
National Pipe Thread (NPT), 316
data, 311
Standards, 312
Natural deterioration, 23
Natural life expectancy, 23
Net operation time, 271–273
Non-bottleneck machines, 17, 274
capacity, 276
wafer capacity chart, 280, 281
Nongraceful deterioration, 121–123

Non-optimal pump operating
temperatures, 225
Notebooks. *See* Team
NPT. *See* National Pipe Thread
Nut-and-bolt assembly, 143
Nuts, 162

O

Occupational Safety and Health
Administration (OSHA), 341
OEE. *See* Overall equipment effectiveness
One-point lessons, 63, 78–82, 89–90,
152–153, 298
creation/distribution/recording
system, 108
number, 96
procedures, 214
system, 173
usage, 136, 144, 205, 207. *See also*
Equipment; Operators
On-hand quantity, 164
Open M-Tags, 78, 181
Open stock parts, 160–163
Operating conditions, 224, 227
Operating procedures, 297
changes, 205
Operating rate, 271, 272
Operating temperatures. *See* Non-optimal
pump operating temperatures
Operation time, 271. *See also* Net
operation time; Valuable operation
time
Operator-accessible machine
locations, 212
Operators, 78, 109. *See also* Production
cleaning standard, changes, 205
goals, 48
one-point lessons, usage, 80
role, elevation, 170–171
Optical switch assembly, 151–152
Order quantity, 164
Organized workplace, 136, 144, 155–157
Originators, 78
O-ring
boss fittings. *See* SAE O-ring boss
fittings
supply source, 163

Oscilloscopes, 139
OSHA. *See* Occupational Safety and
Health Administration
Out-of-spec condition, 57
Overall equipment effectiveness (OEE),
268–271
concept, 269
tracking, 270

P

P minus (P–), 291
P plus (P+), 209
Paperwork, simplification/
minimization, 157
Pareto approach, 258–260. *See also*
Equipment
Pareto charts, 202. *See also* Machine;
Wafer-handling loss
Part life, standard deviation, 122
Part logs, 113, 117, 126–127
Part stock. *See* Controlled part stock;
Open stock part
Partner and Practice system, 136, 144,
154–155, 173
Parts replacement, 222
Parts/tool management. *See* Spare
parts/tool management
PDCA. *See* Plan Do Check Act
Performance
improvement. *See* Machine
plan, 294
process, 293–294
losses, 268
metric, 229
rate, 269, 271, 272
Personnel skills, elevation, 246
Phillips screwdrivers, 333–334
Physical analysis, 184, 196–201. *See also*
Machine
conducting, 196
Pilot team. *See* Total Productive
Maintenance
Pipe adapters, variety. *See* Swivel pipe
adapters
Pipe joint, 140
Pipe swivel adapter, 317

Pipe threads. *See* Female pipe thread; Male pipe thread
  exercises, 319
  fittings, 310–318
  fittings/swivel pipe adaptors, 137
  knowledge. See Maintenance
  machined fitting, 311
  usage. *See* Hoses
Plan Do Check Act (PDCA), 184, 201–204
Plastic tubing, wear, 52
Plastics, 138
PLC. *See* Programmable logic controller
PM. *See* Preventive Maintenance
P-M. *See* Preventive Maintenance
Pneumatic control valves, 193
Pneumatic fittings, 162
Pneumatic system hardware/schematics, 137
Positive-reinforcement program, 294
Power. *See* Single-phase power; Three-phase power
  transmission, 138
Pozidrive screwdrivers, 333–334
Practice, providing. *See* Guided practice
Practiced levels. *See* Machine failures
Precision
  discipline, 179
  documentation, 136, 144–147
  execution. *See* Maintenance plans implementation
  maintenance, 134. *See also* Master precision maintenance
  work, 149
Precision maintenance support tools, 144–169
  learning time, 153–154
Prefailure conditions, correction, 156
Pre-made hose assembly, 314
Pressure gauges, 121
Pressure regulators, 313
Prevention skills. *See* Failure
Preventive Maintenance (PM / P-M), 342. *See also* Scheduled PM
  activity, 121
  analysis, 184, 191–192

  checklists, 113–116, 178, 298
  checksheets, 117
  document change/training process, 207
  evaluations, 178–179
   process, 207
  execution, 179
   Lite, 184, 192–196
  failure analysis, 196
  machine service data, 117
  notification change, 207
  plan, 7, 114, 211. *See also* Production
   identification. *See* Machine
  procedures, 113, 125–126
  programs, 6–7
  routines, 179
  schedules, 113, 116–124
  usage, 161
  visibility system, 173
  visual route maps, 86–87
Process
  controls. *See* Visual process controls
  flowcharts, 108
  quality control, measure, 268
  recipe, 342
Processing machines. *See* Integrated circuit
Product
  cost, 285
  cycle time, 268
  defect rate, 267
  design, 28
  flow. *See* Linear product flow
  move rates, 267
  scrap, 231–232, 268
  time. *See* Actual product time; Basic product time
  yields, 229, 232, 268
Production, 276. *See also* Total production
  engineers, 246
  flow. *See* Visually controlled production flow
  lot box, 88
  lots, 159
  machine, master PM plan, 118–119
  material, color coding, 159
  operators, 75, 246

rates, 267
  service, machine preparation, 70–71
  wafer boxes, color coding, 157–158
Productivity, 202
  analysis, 228–234
  enhancement, 157
  impact. *See* Manufacturing
  improvement. *See* Continuous
      productivity; Machine productivity
      improvement
  methods. *See* Total Productive
      Maintenance
  increase. *See* Factory
  losses, 228–234, 246, 261
  results, 16
  world-class levels, 39
Program safety. *See* Total Productive
      Maintenance
Programmable logic controller (PLC), 341
Prototype solution, implementation,
      201–203
Pull-card station, 88
Pumps, 137
  performance curve, 228
  temperature, measurement, 227
Punishing consequence, 290

Q

Quality. *See* Manufactured product
  checks, 113, 127–128
  component, 221
  increase, 157
  losses, 268
  maintenance analysis, 218–222
  metric, 219–220
  monitor checklist. *See* Agilent
  product, 28, 232
  production. *See* Total quality
      production
  rate, 269, 271, 272
Quantity. *See* On-hand quantity; Order
      quantity
Quick-connect fitting, 216
Quick-connect fittings, 137

R

R minus (R–), 290

R plus (R+), 290
Radio frequency (RF), 342
Ranking systems, changes, 108
Reactive maintenance, 29
Rebuild and swap technique, 178–179
Redesigns. *See* Machine redesigns
Refrigeration systems, 139
Replacement interval, 21
Replacement maintenance, 18
  scheduled, 21
Replacement parts, 125
  numbers, 113
Restoration. *See* Abnormalities;
      Equipment restoration; Machine
maintenance schedule, creation, 194
Results, evaluation, 203
Reward systems, change, 108
RF. *See* Radio frequency
Robot. *See* Disassembled robot; Wafer-
      handling robot
  assembly, 152
  effect. *See* Wafer
  misalignment, reasons, 186
Robot base
  locating, 147
  mounting. *See* Wafer-Handler
      Rebuild Cart
  orientation, 148
Rod ends, 224
Root cause. *See* Equipment
  analysis, 201, 202
  correction, 203
Rotation schedule, 223
Route maps, 86, 108. *See also* Visual
      route maps
  order, 86
Routine maintenance, 62

S

SAE fittings, 322–324
  exercises, 324
  screw, 323
  sizes, 324
SAE grades, 335
SAE O-ring boss fittings, 137, 322, 323
Safety. *See* Sort Stabilize Shine
      Standardize Sustain Safety Security

audits, 68
behavior observers, certification, 68
committee. *See* Behavior
concerns, 89, 95
guards, 52
incidents, review, 68
interval, 122
management systems, 64
panels, 101
program. See Agilent
program data/results, 69
Scatter diagrams, 202
Scheduled monitors, 121
Scheduled PM, 181
Scheduled time, 271–273
Schedules. *See* Preventive Maintenance
design. *See* Maintenance
Scheduling computer, 102, 103
Scrap. *See* Product
measurements, 232
Screwdrivers. *See* Phillips screwdrivers;
Pozidrive screwdrivers
comparison, 141, 333
Sealants, 138
Seals, 137. *See also* Dynamic seals;
Static seals
Security. *See* Employee; Sort Stabilize Shine
Standardize Sustain Safety Security
measures. *See* Factory
Service data. *See* Preventive
Maintenance
Service interval, choice, 20
Set screws, 138
Shape, frequency interaction, 86
Shift ownership, 86
Shine. *See* Sort Stabilize Shine
Standardize Sustain Safety Security
Shock absorbers, 224
Shoulder assembly, 149
Single-phase power, 138
Sizing tools, 163
Skills
elevation, 170–171
training. *See* Technical skill training
Slit valve, 193
assembly, 195
Snap rings, 138

Socket head bolts, 142, 335–336
types, 336
Software backups, 156
Sort. *See* Sort Stabilize Shine Standardize
Sustain Safety Security
Sort Stabilize Shine (3S), 156
Sort Stabilize Shine Standardize
Sustain (5S)
locations, 212, 214
practices, 156
principles, 114
process, 168
shopfloor organization system, 173
workplace, 155–156
Sort Stabilize Shine Standardize Sustain
Safety Security (7S) foundation,
155–157
Spare parts/tool management, 136, 144,
160–169
Speed losses, 233, 234
Splines, 138
Sporadic loss, 22
SS bolt, 187
SSIP rules, 291–293
Stabilize. *See* Sort Stabilize Shine
Standardize Sustain Safety Security
Standardize. *See* Sort Stabilize Shine
Standardize Sustain Safety Security
Standards, setting. *See* Machine
Start-up level, 181–182. *See also* Machine
failures
Static charge sensor, 243
Static seals, 137
Steering box, 224
Stock parts. *See* Open stock parts
Storage
areas, identification, 157
elevator, 193
Straight thread-O ring boss, 322–324
exercises, 324
Strategy, 251
Struts, 224
Subassembly priority, 262–264
Subjects, breakdown, 11
Success criteria, 203
Support tools. *See* Precision maintenance
support tools

Suspension parts, 224
Sustain. *See* Sort Stabilize Shine
    Standardize Sustain Safety Security
Swagelok fittings, re-assembly
    sequence, 329
Swagelok internal tubing support, 327
Swagelok tube fittings, 137, 326–329
    inspection. *See* Assembled Swagelok
    tube fittings
Swap technique. *See* Rebuild and swap
    technique
Switches
    assembly. *See* Optical switch assembly
    improper settings, 52
Swivel adapter. *See* Mating swivel
    adapter; Pipe swivel adapter
Swivel pipe adapters, variety, 318
System Biweekly Inspection PM,
    114, 115
System factors, identification, 193
System Lubrication PM, 114

T
Tactical plan, 251. *See also* Agilent
    manufacturing
Tagout, 341
Taper-locks, 138
Taste test, 52
Team. *See* Action teams; Cross-
    departmental teams; Hazardous
    Material; Improvement; Total
    Productive Maintenance
    access, 70
    activities, advancing, 96–107
    activity
        boards, 63, 89, 295. *See also* Agilent
        data, 95
        improvement ideas, 89
    approach. *See* Machine productivity
        improvement
    challenge, 37
    charter, 89
    chartering, 37
    colors, 87
    concept. *See* Total Productive
        Maintenance
    goal, 297

initiating. *See* Total Productive
    Maintenance
    leader, response, 292
    meetings, 96
    members, 89, 292
        contingent consequences delivery, 37
        training, 37
        usage. *See* Machine; Machine
            standards setting
    metrics, trend charts, 89
    notebooks, 63, 89
    ownership, 86
    performance charts, 89
Technical minicourses, preparation,
    139–143
Technical skill training, 136
Technicians
    decisions, 130
    goals, 48
Technician-scheduled maintenance plans,
    205
Technology
    acquisition, 140
    lessons. *See* Generic technology
        lessons
Technology Quality Responsiveness
    Delivery Cost (TQRDC), 7–8
Teflon tape, usage, 140, 312
Tell-tale signs, 121
Temperature. *See* Non-optimal pump
    operating temperatures
    gauges, 121
    measurement. *See* Pumps
Test bolts, failure, 188
Test pieces, assembly/fabrication, 187
Text gauges, 235, 240
Text maintenance procedure, 126
Thread-locking devices, 137
Threads, 137. *See also* Aluminum thread
    size. *See* JIC thread size
    strength, 143
    usage, 140
Three Lists, 63, 68–74, 89
    examples. *See* Agilent
Three-phase power, 138
Throughput, 202
Tic marks, 117

Tickler file, 117
Tie wraps, 75
Time to learn. *See* Learning
Time-based maintenance, 62
    plans, creation, 192
Tires, balance, 223
Tool management. *See* Spare parts/tool
    management
Toolkit. *See* Equipment restoration
Tools. *See* Failure; Maintenance
Total production, 271
Total production ratio (TPR), 270–274
Total Productive Maintenance (TPM)
    action teams, 184
        deployment, 36
        implementation, 36–38
    activities, 34, 177, 246, 290, 302, 306
    activity checklist, 97
    approach, 260–264
    attack, process. *See* Machine loss
    changes, sustaining, 299
    culture change, 9–11
    derivation terms, 271
    focus. *See* Machine components
    fundamentals, 3
    gap analysis, 302
    implementation, 262
    improvement methods, 16
    master plan, 43
    methods, application, 19
    pilot team, 34–36, 96, 253
    principles, 5–7
    productivity improvement methods, 37
    program, 171
        safety, 63–74
    pyramid. *See* Chronic conditions
step-by-step implementation, 41–43
steps
        activities, master checklist, 89
        focusing, 261–264
        knowledge, 39
        summary, 247–248
success, keys, 39
teams, 135, 171, 206, 246, 247
        concept, 38
        initiating, 95–96
tools, usage, 37

usage. *See* Machine productivity
    improvement
Total quality production, 271
TPM. *See* Total Productive Maintenance
TPR. *See* Total production ratio
TQRDC. *See* Technology Quality
    Responsiveness Delivery Cost
Training
    activities, 214
    certification, 68
    delivery, 143
    guidelines, 143
    material, 214
    plans, creation, 192, 194
    providing, 136
    support, 246
Trend charts, 202. *See also* Condition-
    measurement trend charts;
    Machine; Team
Trouble lists. *See* Failure
Tubing, 137

U

Ulta-Torr vacuum fittings, 137
Ultrasonic analysis, 234, 236
Ultrasonic leak detector, 236
Unanswered questions, 95
Underdeveloped maintenance
    plans, 20
Use-based maintenance, 62
    plans, creation, 192

V

Vacuum bellows assembly, 190
Vacuum pump, 226
    distribution manifold/cooling water
        thermostat, 198
    types, 216
Vacuum systems, 139
Valuable operation time, 271–273
Valves, 137. *See also* I/O valve; Pneumatic
    control valves
VCO fittings, 137
VCR fittings, 137
Vibration, 59. *See also* Machine
    analysis, 234, 235
Video-imaging analysis, 235, 238

Visibility system. *See* Preventive
    Maintenance
Vision, 251. *See also* Agilent
    manufacturing
Visual controls, 63, 83–88. *See also*
    Flow meter
  minor defects, 57
Visual factory floor. *See* Agilent
Visual inspections, 51
Visual machine controls, 205
Visual maintenance documentation
    system, 173
Visual process controls, 205
Visual route maps. *See* Preventive
    Maintenance
  locators, 85
Visual workplace, creation, 157–159
Visually controlled input station, 231
Visually controlled inspection routes,
    84–86
Visually controlled production flow,
    87–88
Volt/ohm meters, 138

W

Wafer. *See* Integrated circuit
  anneal, 71
  boxes, color coding. *See* Production
  capacity chart. *See* Bottleneck
    machines; Non-bottleneck machines
  cassette, 67
    minor abnormalities, 240
    removal, 65
    vertical position, 66
  damage, 186
  fall, 185
  lot, 342
  processing, 233
  scratches, robot effect, 186
  transfer device, 66
Wafer-Handler Rebuild Cart, Robot Base
    mounting, 147–152
Wafer-handling loss, Pareto charts, 259
Wafer-handling robot, 146, 193
Wafer-handling system, 262
Washers. *See* Lock washers
Waste, recognition, 157

Wastes resources, guessing, 188–189
Water-quality analysis, 235, 238–239
Weak machine design, 20
  changes, 205
Wear particle analysis, 235–237
Winding insulation properties, 239
WIP. *See* Work in progress
Wires, securing, 185
Wiring insulation, 52
Work areas, 47
Work habits. *See* Employee
Work in progress (WIP), 342
  cost, 280
  racks, organization/labeling, 159
Work order, 181
  generation, 180
Work routine. *See* Day-to-day work
    routine
Working environment, 64–67
Working time, 271–273
Workplace. *See* Organized workplace
  cleanliness, 156, 157
  creation. *See* Visual workplace
Wrists
  neutral position, 65, 66
  rotation, 66

Y

Y connections, 138

Books from Productivity Press

Productivity Press publishes books that empower individuals and companies to achieve excellence in quality, productivity, and the creative involvement of all employees. Through steadfast efforts to support the vision and strategy of continuous improvement, Productivity Press delivers today's leading-edge tools and techniques gathered directly from industry leaders around the world. Call toll-free (800) 394-6868 for our free catalog.

Pillars of the Visual Workplace
The Sourcebook for 5S Implementation

Hiroyuki Hirano

In this important sourcebook, JIT expert Hiroyuki Hirano provides the most vital information available on the visual workplace. He describes the 5S's: in Japanese they are seiri, seiton, seiso, seiketsu, and shitsuke (which translate as organization, orderliness, cleanliness, standardized cleanup, and discipline). Hirano discusses how the 5S theory fosters efficiency, maintenance, and continuous improvement in all areas of the company, from the plant floor to the sales office. This book includes case materials, graphic illustrations, and photographs.
ISBN 1-56327-047-1 / 377 pages, illustrated / $85.00

20 Keys to Workplace Improvement (Revised Edition)

Iwao Kobayashi

The 20 Keys system does more than just bring together twenty of the world's top manufacturing improvement approaches—it integrates these individual methods into a closely interrelated system for revolutionizing every aspect of your manufacturing organization. This revised edition of Kobayashi's bestseller amplifies the synergistic power of raising the levels of all these critical areas simultaneously. The new edition presents upgraded criteria for the five-level scoring system in most of the 20 Keys, supporting your progress toward becoming not only best in your industry but best in the world.
ISBN 1-56327-109-5/ 302 pages / $50.00

DOE Simplified
Practical Tools for Effective Experimentation

Mark J. Anderson and Patrick J. Whitcomb

Design of Experiment (DOE) is a planned approach for determining cause and effect relationships. The tool can be applied to any process with measurable input and outputs. *DOE Simplified* is a comprehensive new introductory text that is geared for readers with minimal statistical. Filled with fun anecdotes and sidebars, the text cuts through the complexities of this powerful improvement tool.
ISBN 1-56327-225-3 / 250 pages / $39.95

Productivity Press, Dept. BK, P.O. Box 13390, Portland, OR 97213-0390
Telephone: 1-800-394-6868   Fax: 1-800-394-6286

Eliminating Minor Stoppages on Automated Lines

Kikuo Suehiro

Stoppages of automated equipment lines severely affect productivity, cost, and lead time. Such losses make decreasing the number of stoppages a crucial element of TPM. Kikuo Suehiro has helped companies such as Hitachi achieve unprecedented reduction in the number of minor stoppages. In this explicitly detailed book, he presents a scientific approach to determining the causes of stoppages and the actions that can be taken to diminish their occurrence.
ISBN 0-915299-70-4 / 243 pages / $50.00

Equipment Planning for TPM
Maintenance Prevention Design

Fumio Gotoh

This practical book for design engineers, maintenance technicians, and manufacturing managers details a systematic approach to the improvement of equipment development and design and product manufacturing. The author analyzes five basic conditions for factory equipment of the future: development, reliability, economics, availability, and maintainability. The book's revolutionary concepts of equipment design and development enables managers to reduce equipment development time, balance maintenance and equipment planning and improvement, and improve quality production equipment.
ISBN 0-915299-77-1 / 337 pages / $65.00

Fast Track to Waste-Free Manufacturing:
Straight Talk from a Plant Manager

John W. Davis

Batch, or mass, manufacturing is still the preferred system of production for most U.S.-based industry. But to survive, let alone become globally competitive, companies will have to put aside their old habitual mass manufacturing paradigms and completely change their existing system of production. In *Fast Track to Waste-Free Manufacturing: Straight Talk from a Plant Manager*, John Davis details a new and proven system called Waste-Free Manufacturing (WFM) that rapidly deploys the lean process. He covers nearly every aspect of the lean revolution and provides essential tools and techniques you will need to implement WFM. Drawing from more than 30 years of manufacturing experience, John Davis gives you tools and techniques for eliminating anything that cannot be clearly established as value added.
ISBN: 1-56327-212-1 / 425 pages / $45.00

Productivity Press, Dept. BK, P.O. Box 13390, Portland, OR 97213-0390
Telephone: 1-800-394-6868   Fax: 1-800-394-6286

Handbook for Productivity Measurement and Improvement

William F. Christopher and Carl G. Thor, eds.

An unparalleled resource! In over 100 chapters, nearly 80 front-runners in the quality movement reveal the evolving theory and specific practices of world class organizations. Spanning a wide variety of industries and business sectors, they discuss quality and productivity in manufacturing, service industries, profit centers, administration, nonprofit and government institutions, health care and education. Contributors include Robert C. Camp, Peter F. Drucker, Jay W. Forrester, Joseph M. Juran, Robert S. Kaplan, John W. Kendrick, Yasuhiro Monden, and Lester C. Thurow. Comprehensive in scope and organized for easy reference, this compendium belongs in every company and academic institution concerned with business and industrial viability.
ISBN 1-56327-007-2 / 1344 pages / $90.00

Handbook of Quality Tools
The Japanese Approach

Tetsuichi Asaka and Kazuo Ozeki (eds.)

The Japanese have stunned the world by their ability to produce top quality products at competitive prices. This comprehensive teaching manual, which includes the seven traditional and five newer QC tools, explains each tool, why it's useful, and how to construct and use it. It's a perfect training aid, as well as a hands-on reference book, for supervisors, foremen, and/or team leaders. Accessible to everyone in your organization, dealing with both management and shop floor how-to's, you'll find it an indispensable tool in your quest for quality. Information is presented in easy-to-grasp language, with step-by-step instructions, illustrations, and examples of each tool.
ISBN 1-56327-138-9 / 315 pages / $30.00

Introduction to Implementing TPM
The North American Experience

Charles J. Robinson and Andrew P. Ginder

The authors document an approach to TPM planning and deployment that modifies the Japan Institute of Plant Maintenance 12-step process to accommodate the experiences of North American plants. They include details and advice on specific deployment steps, OEE calculation methodology, and autonomous maintenance deployment. This book shows how to make TPM work in unionized plants and how to position TPM to support and complement other strategic manufacturing improvement initiatives. More than just an implementation guide, it's actually a testimonial of proven TPM success in North American companies through the adoption of "best in class" manufacturing practices.
ISBN 1-56327-087-0 / 224 pages / $45.00

Productivity Press, Dept. BK, P.O. Box 13390, Portland, OR 97213-0390
Telephone: 1-800-394-6868   Fax: 1-800-394-6286

Quick Response Manufacturing
A Companywide Approach to Reducing Lead Times

Rajan Suri

Quick Response Manufacturing (QRM) is an expansion of time-based competition (TBC) strategies, which use speed for a competitive advantage. Essentially, QRM stems from a single principle: to reduce lead times. But unlike other time-based competition strategies, QRM is an approach for the entire organization, from the front desk to the shop floor, from purchasing to sales. In order to truly succeed with speed-based competition, you must adopt the approach *throughout* the organization.
ISBN 1-56327-201-6/ 560 pages / $50.00

TPM for America
What It Is and Why You Need It

Herbert R. Steinbacher and Norma L. Steinbacher

As much as 15 to 40 percent of manufacturing costs are attributable to maintenance. With a fully implemented TPM program, your company can eradicate all but a fraction of these costs. Co-written by an American TPM practitioner and an experienced educator, this book gives a convincing account of why American companies must adopt TPM if they are to successfully compete in world markets. Includes examples from leading American companies showing how TPM has changed them into more efficient and productive organizations.
ISBN 1-56327-044-7 / 169 pages / $25.00

TPM for Workshop Leaders

Kunio Shirose

A top TPM consultant in Japan, Kunio Shirose describes the problems that TPM group leaders are likely to experience and the improvements in quality and vast cost savings you should expect to achieve. In this non-technical overview of TPM, he incorporates cartoons and graphics to convey the hands-on leadership issues of TPM implementation. Case studies and realistic examples reinforce Shirose's ideas on training and managing equipment operators in the care of their equipment.
ISBN 0-915299-92-5 / 164 pages / $40.00

TPM in Process Industries

Tokutaro Suzuki (ed.)

Process industries have a particularly urgent need for collaborative equipment management systems like TPM that can absolutely guarantee safe, stable operation. In *TPM in Process Industries*, top consultants from JIPM (Japan Institute of Plant Maintenance) document approaches to implementing TPM in process industries. They focus on the process environment and equipment issues such as process loss structure and calculation, autonomous maintenance, equipment and process improvement, and quality maintenance. Must reading for any manager in the process industry.
ISBN 1-56327-036-6 / 400 pages / $85.00

Productivity Press, Dept. BK, P.O. Box 13390, Portland, OR 97213-0390
Telephone: 1-800-394-6868   Fax: 1-800-394-6286

Training for TPM
A Manufacturing Success Story

Nachi-Fujikoshi (ed.)

A detailed case study of TPM implementation at a world-class manufacturer of bearings, precision machine tools, dies, industrial equipment, and robots. The book details how the company trained managers and workers and shows the improvements they achieved in reducing breakdowns and defects while revitalizing the workforce. In just 2½ years the company was awarded Japan's prestigious PM Prize for its program. Here is a detailed account of their improvement activities—and an impressive model for yours.
ISBN 0-915299-34-8 / 274 pages / $50.00

Uptime
Strategies for Excellence in Maintenance Management

John Dixon Campbell

Campbell outlines a blueprint for a world class maintenance program by examining, piece by piece, its essential elements—leadership (strategy and management), control (data management, measures, tactics, planning and scheduling), continuous improvement (RCM and TPM), and quantum leaps (process reengineering). He explains each element in detail, using simple language and practical examples from a wide range of industries. This book is for every manager who needs to see the "big picture" of maintenance management. In addition to maintenance, engineering, and manufacturing managers, all business managers will benefit from this comprehensive and realistic approach to improving asset performance.
ISBN 1-56327-053-6 / 204 pages / $35.00

Productivity Press, Dept. BK, P.O. Box 13390, Portland, OR 97213-0390
Telephone: 1-800-394-6868   Fax: 1-800-394-6286

TO ORDER: Write, phone, or fax Productivity Press, Dept. BK, P.O. Box 13390, Portland, OR 97213-0390, phone 1-800-394-6868, fax 1-800-394-6286. Outside the U.S. phone (503) 235-0600; fax (503) 235-0909. Send check or charge to your credit card (American Express, Visa, MasterCard accepted).

U.S. ORDERS: Add $5 shipping for first book, $2 each additional for UPS surface delivery. Add $5 for each AV program containing 1 or 2 tapes; add $15 for each AV program containing 3 or more tapes. We offer attractive quantity discounts for bulk purchases of individual or mixed titles; call for more information.

ORDER BY E-MAIL: Order 24 hours a day from anywhere in the world. Use either address:
To order: **info@productivityinc.com**
To view the online catalog and/or order: **http://www.productivityinc.com/**

QUANTITY DISCOUNTS: For information on quantity discounts, please contact our sales department.

INTERNATIONAL ORDERS: Write, phone, or fax for quote and indicate shipping method desired. For international callers, the telephone number is 503-235-0600 and the fax number is 503-235-0909. Prepayment in U.S. dollars must accompany your order (checks must be drawn on U.S. banks). When quote is returned with payment, your order will be shipped promptly by the method requested.

NOTE: *Prices are in U.S. dollars and are subject to change without notice.*

About the Shopfloor Series

Put powerful and proven improvement tools in the hands of your entire workforce!

Progressive shopfloor improvement techniques are imperative for manufacturers who want to stay competitive and to achieve world class excellence. And it's the comprehensive education of all shopfloor workers that ensures full participation and success when implementing new programs. The Shopfloor Series books make practical information accessible to everyone by presenting major concepts and tools in simple, clear language and at a reading level that has been adjusted for operators by skilled instructional designers. One main idea is presented every two to four pages so that the book can be picked up and put down easily. Each chapter begins with an overview and ends with a summary section. Helpful illustrations are used throughout.

Books currently in the Shopfloor Series include:

5S for Operators
5 Pillars of the Visual Workplace
The Productivity Press Development Team
ISBN 1-56327-123-0 / 133 pages
$25.00

Quick Changeover for Operators
The SMED System
The Productivity Press Development Team
ISBN 1-56327-125-7 / 93 pages
$25.00

Mistake-Proofing for Operators
The Productivity Press Development Team
ISBN 1-56327-127-3 / 93 pages
$25.00

Just-In-Time for Operators
The Productivity Press Development Team
ISBN 1-56327-133-8 / 84 pages
$25.00

TPM for Supervisors
The Productivity Press Development Team
ISBN 1-56327-161-3 / 96 pages
$25.00

TPM Team Guide
Kunio Shirose
ISBN 1-56327-079-X / 175 pages
$25.00

Autonomous Maintenance
Japan Institute of Plant Maintenance
ISBN 1-56327-082-X / 138 pages
$25.00

Focused Equipment Improvement for TPM Teams
Japan Institute of Plant Maintenance
ISBN 1-56327-081-1 / 138 pages
$25.00

TPM for Every Operator
Japan Institute of Plant Maintenance
ISBN 1-56327-080-3 / 136 pages
$25.00

OEE for Operators
Overall Equipment Effectiveness
The Productivity Press Development Team
ISBN 1-56327-221-0 / 96 pages
$25.00

Cellular Manufacturing
The Productivity Press Development Team
ISBN 1-56327-213-X / 96 pages
$25.00

Productivity Press, Dept. BK, P.O. Box 13390, Portland, OR 97213-0390
Telephone: 1-800-394-6868   Fax: 1-800-394-6286

Productivity, Inc. Consulting, Training, Workshops, and Conferences

EDUCATION...IMPLEMENTATION...RESULTS

Productivity, Inc. is the leading American consulting, training, and publishing company focusing on delivering improvement technology to the global manufacturing industry.

Productivity prides itself on delivering today's leading performance improvement tools and methodologies to enhance rapid, ongoing, measurable results. Whether you need assistance with long-term planning or focused, results-driven training, Productivity's world-class consultants can enhance your pursuit of competitive advantage. In concert with your management team, Productivity will focus on implementing the principles of Value-Adding Management, Total Quality Management, Just-in-Time, and Total Productive Maintenance. Each approach is supported by Productivity's wide array of team-based tools: Standardization, One-Piece Flow, Hoshin Planning, Quick Changeover, Mistake-Proofing, Kanban, Problem Solving with CEDAC, Visual Workplace, Visual Office, Autonomous Maintenance, Overall Equipment Effectiveness, Design of Experiments, Quality Function Deployment, Ergonomics, and more! And, based on continuing research, Productivity expands its offering every year.

Productivity's conferences provide an excellent opportunity to interact with the best of the best. Each year our national conferences bring together the leading practitioners of world-class, high-performance strategies. Our workshops, forums, plant tours, and master series are scheduled throughout the U.S. to provide the opportunity for continuous improvement in key areas of lean management and production.

Productivity, Inc. is known for significant improvement on the shop floor and the bottom line. Through years of repeat business, an expanding and loyal client base continues to recommend Productivity to their colleagues. Contact us at 1-800-394-6868 to learn how we can tailor our services to fit your needs.